# GENERAL CHEMISTRY I
## Laboratory Manual

## Chem 131L

*Third Edition*

Department of Chemistry
Towson University

**General Chemistry I Laboratory Manual, Third Edition – CHEM 131L**

Copyright © 2017 by the Department of Chemistry, Towson University

All rights reserved. No part of this publication may be reproduced or transmitted in any form or by any means, electronic or mechanical, including photocopying, recording, or any information storage and retrieval system, without the written permission of the publisher.

Requests for permission to make copies of any part of the work should be mailed to:

Permissions Department
Academx Publishing Services, Inc.
P.O. Box 208
Sagamore Beach, MA 02562
http://www.academx.com

Printed in the United States of America

ISBN-13: 978-1-68284-299-7
ISBN-10: 1-68284-299-1

THIS BOOK IS PRINTED ON RECYCLED PAPER 

# Table of Contents

# Introduction

Chemistry is an experimental science and its study requires appropriate laboratory work. The Department of Chemistry faculty and staff believe that your understanding of chemistry will be deeper if a significant and sometimes challenging laboratory experience is an integral part of the General Chemistry curriculum.

We have designed the laboratory component of the course to develop your observational, manipulative, critical thinking, and communications skills while reinforcing your understanding of important chemical concepts.

The initial part of the laboratory program introduces you to the basics of recording and reporting laboratory results. Your ability to communicate effectively what you have done in the laboratory depends to a large extent on your understanding of the experiment and the chemical principles illustrated.

The first few experiments build your observational and manipulative skills through practice in accurately and efficiently using common laboratory equipment. The focus of these experiments is to develop your ability to measure, both qualitatively and quantitatively, the physical properties of substances. These are basic skills you will build upon in later experiments, as well as later courses in the chemistry program.

This Laboratory Manual is the collaborative work of the faculty and staff of the Department of Chemistry. Some experiments in the manual were developed with the support of the National Science Foundation. We gratefully acknowledge NSF CCLI Adapt and Adopt Grant DUE 9950952: "Improving Student Learning in Chemistry; Building on the New Traditions Project."

We hope your laboratory experience is rewarding and interesting, and that it helps you in meeting your goals in General Chemistry.

# Record Keeping in the Laboratory

A record must be kept describing what you do in the laboratory. This record is kept in the form of a laboratory notebook. The discussion below describes how to format a laboratory notebook and record the information obtained in the laboratory.

## A. The Laboratory Notebook

The laboratory notebook is the reference documenting what occurs in experiments. It is the source book for the preparation of reports, and is a permanent record that may be consulted after even several years have passed. In all experimental work, it is standard practice to record *everything* relevant to the experiment (data, observations, notes and comments, literature information, calculations and graphs) *in ink*, *directly* into a *bound* notebook with *numbered* pages. It should *honestly* represent what you did while working in the lab, including any errors that were made.

There are three main requirements that must be met when entering your work in the laboratory into your notebook:

1. The record should be *complete*.

2. The record must be *intelligible* to others skilled in chemistry.

3. The record should be *easy to find* in the notebook on short notice.

You are required to maintain a notebook with sequentially numbered pages, capable of making duplicate copies. These notebooks may be purchased from the University Store or from the Stu-dent Affiliates of the American Chemical Society (SAACS). The general format for keeping a laboratory notebook is given below.

## B. Notebook Format

The back cover of the notebook has a flap which is to be placed under the duplicate page to prevent you from making more than one duplicate page. *Before you begin writing in your lab notebook, be sure the flap is in place.*

The first two pages of the notebook are titled **Table of Contents**. You should keep this up-to-date throughout the semester.

The records of your laboratory experiments begin on page 1. Each new experiment must begin on a new page, and information from two or more different experiments should not be placed on the same notebook page. On the first page for each experiment, fill in all of the required information at the top of the page (in the blue boxes). The bottom of each page shows your signature, along with the signature of another person who witnessed the work being done in the laboratory, together with the date. Your "witness" will be your laboratory instructor.

The body of any page in the notebook (see a typical notebook page below) contains one or more of the following: **Laboratory Lecture Notes, Experimental Procedure/Data & Observations**, and **Calculations**. Separate these sections clearly with these headings. These are discussed separately below.

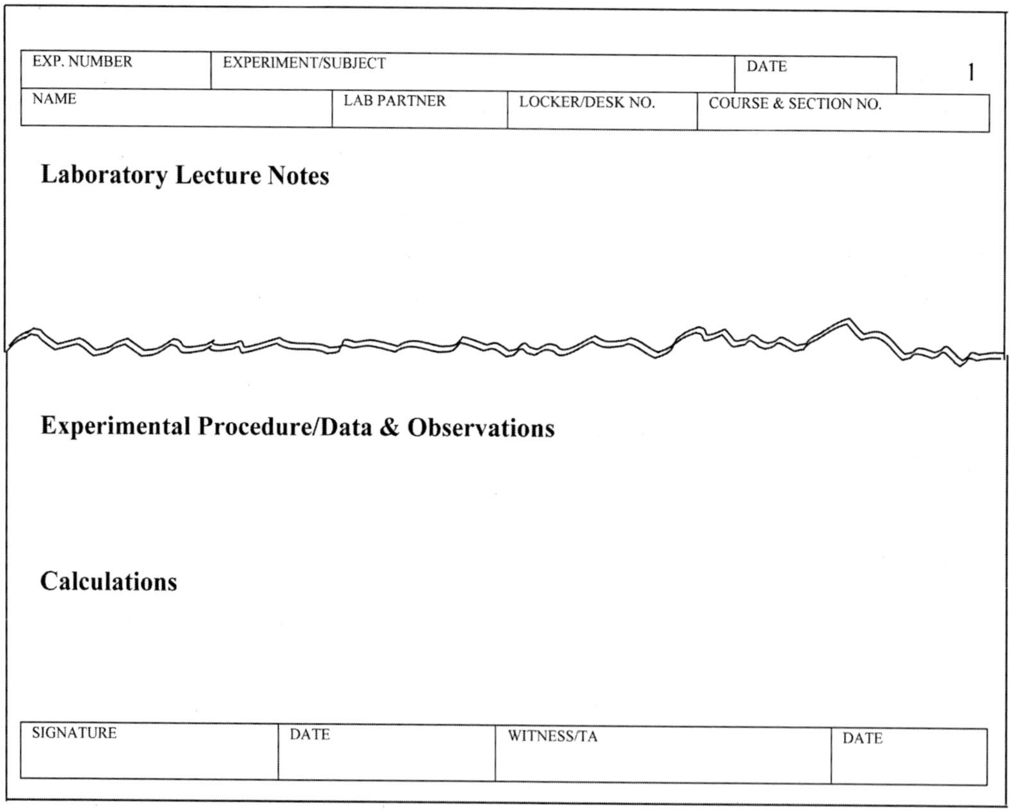

| EXP. NUMBER | EXPERIMENT/SUBJECT | | | DATE |
|---|---|---|---|---|
| NAME | LAB PARTNER | LOCKER/DESK NO. | COURSE & SECTION NO. | |

1

**Laboratory Lecture Notes**

**Experimental Procedure/Data & Observations**

**Calculations**

| SIGNATURE | DATE | WITNESS/TA | DATE |
|---|---|---|---|

**Laboratory Lecture Notes:** Your instructor will provide an introduction to each experiment prior to the laboratory session. A typical introductory laboratory lecture would include a discussion of the theory or principles of the experiment, a review of laboratory procedures, and a discussion of calculations. Important changes or modifications of procedures, helpful hints and safety issues are often presented. There is also likely to be a discussion of the format, due date, and contents of the post-lab assignment for that experiment.

**Experimental Procedure/Data & Observations:** This section begins with a summary of the procedure, or a reference to the source of the procedure. Data and observations are recorded here as they are collected. *Data* are quantitative values for measurements of mass, volume, temperature, concentrations, etc. The data should be carefully organized and labeled so that needed information can be found and understood easily. The use of tables is particularly helpful in keeping this section well-organized. *Observations* are qualitative items, and include such things as the color change observed when solutions were mixed, effects observed as the reaction occurs (i.e. volume change, production of gases, etc.), and any other details that might be relevant to an understanding of what you did or observed during the course of the experiment.

**Calculations:** Calculations needed to achieve the desired result of the experiment are contained in this section. *It is recommended that these calculations be completed* ***before leaving*** *the laboratory whenever possible, so that the quality of the data may be assessed.* A poor or unexpected result should cause you to think about possible sources of error, and may make it necessary for you to repeat one or more portions of the experiment before leaving the laboratory.

## C. Guidelines for Entries in the Laboratory Notebook

The following are some guidelines to, and generally accepted practices in, keeping a notebook.

- All entries must be written in non-erasable ink (blue or black), be legible, and well-spaced from each other. The duplicate copies must be legible as well.

- Record all data and observations *directly* into the notebook. Data should never be recorded on loose scraps of paper.

- Entries should be clearly identified with labels, stating what the entry represents as well as its units.

- Never erase an entry or use white-out. Rather, use a single line to neatly cross it out (don't obliterate it!) and briefly explain, in the margin, why it is to be voided.

- *The original pages should* ***never*** *be torn from the notebook.* All entries in the lab notebook are considered a *permanent* record.

- Although entries should be complete, neat and legible, you should not waste time trying to make your notebook look perfect. It is a "work in progress," and, as such, is not expected to be flawless.

- Be sure to record all *raw* data and not merely the calculated result. 'Raw data' means data obtained directly from the measuring device and not derived by any calculations. For example, a student collected the data below when weighing out three ~ 2.5 g quantities of sodium chloride. The mass of the empty beaker and the beaker + NaCl are 'raw data', and the mass of the NaCl was calculated from the difference.

| **Trial Number** | Mass of empty beaker (g) | Mass of beaker + NaCl (g) | Mass of NaCl (g) |
|---|---|---|---|
| 1 | 10.305 | 12.846 | 2.541 |
| 2 | 11.026 | 13.414 | 2.388 |
| 3 | 10.754 | 13.195 | 2.441 |

Experiment 1

# Measurements in the Laboratory

## PURPOSE

To practice performing several laboratory measurement techniques and understand significant figures related to different measuring devices.

## INTRODUCTION

Making measurements is a primary method that scientists use to learn about a sample of interest. Therefore, learning when and how to properly use the different measuring devices available is an important first step for working in a laboratory environment. Equally important is understanding the uncertainty associated with each measuring device, i.e. significant figures in a measurement.

## PROCEDURE

**Special Equipment:** Four-sided meter stick and blocks

**Laboratory Safety:**

- Be sure that your safety goggles are in place when working in the laboratory

There are four stations in different locations of the laboratory. Each pair of students must rotate through all four stations so that each set of measurements is made by all groups. Directions will be available at each station detailing the type of measurement and calculations required. Working with your partner, you may begin at any station. Although you will work with a partner, you must keep an individual record of the experiment. All data, observations and calculations you make during the experiment should be recorded in ink directly in your laboratory notebook.

## Experiment 2

# How Can the Density of a Substance Be Determined with Accuracy and Precision?

## PURPOSE

To apply the techniques of volumetric measurements to investigate the accuracy and precision of glassware and make use of this knowledge to plan and execute the determination of the density of a substance.

## INTRODUCTION

The density of a substance can be determined in the laboratory by measuring the mass and volume of a sample. In general, measuring the mass can be easily accomplished with high precision and accuracy using a balance. On the other hand, measuring the *volume* to the same degree of accuracy and precision is more difficult. The accuracy and precision of the measurement depends on the apparatus used.

In the first part of this experiment, you will use three different pieces of glassware to measure a volume of 10 mL of water. You will then compare the accuracy and precision of these measurements to enable you to choose the best method for volume determination. In the second part of the experiment, you will use your chosen method for measuring volume to determine the density of an unknown liquid. Before you perform this experiment you should study the appropriate sections in Appendix 5, titled "Laboratory Equipment and Techniques," and Appendix 3, titled "Accuracy, Precision, and Experimental Error."

## PRE-LABORATORY ASSIGNMENT

Before you come to lab, plan and prepare a table in your laboratory notebook to record your raw data in an organized manner. Do this for Part A only. The remaining data tables will be prepared during lab time.

## PROCEDURE

**Special Equipment:** 10-mL volumetric pipet, thermometer, 25-mL graduated cylinder, 50-mL beaker.

**Materials:** tap water, "unknown" sample (de-carbonated Pepsi™; regular or Diet)

**Laboratory Safety:**

- Be sure that your safety goggles are in place when working in the laboratory.

## Part A: Using a Beaker to Measure Water Volume

In this section, you will test the ability of one 50-mL beaker to reproducibly contain a given volume of water.

Obtain at least 200 mL of tap water in a clean beaker and allow it to equilibrate for several minutes to ambient temperature, recording the temperature to ± 0.1°C. Weigh one clean, dry, 50-mL beaker and record its mass to mg (0.000 g). Add the water to the 50-mL beaker until it is level with the 10 mL mark on the beaker, and record the new mass. Empty the beaker, dry it with a paper towel, and repeat the process above for a minimum of three trials.

## Part B: Using a Graduated Cylinder to Measure Water Volume

In this section, you will test the ability of a 25-mL graduated cylinder to reproducibly deliver a given volume of water into a beaker.

Weigh three clean, dry, small beakers and record their masses to mg (0.000 g). Use a 25-mL graduated cylinder to deliver 10 mL of the water obtained in Part A to each beaker. Record the new mass of each beaker with water. A minimum of three trials is required.

## Part C: Using a Volumetric Pipet to Measure Water Volume

In this section, you will test your ability to use the pipet to reproducibly deliver a known volume of water into a beaker.

Pipets are designed to deliver known volumes of liquid from one container to another when filled to the calibration mark. Volumetric or "transfer" pipets deliver a single, fixed volume. Your instructor will demonstrate the preferred technique for rinsing and using the 10-mL pipet to reproducibly deliver 10.00 mL of a liquid into a container. Use of a volumetric pipet is described in Appendix 5, Part e.

Weigh three clean, dry, small beakers and record their masses to mg (0.000 g). Use a 10-mL pipet to deliver 10 mL of the water obtained in Part A to each beaker. Record the new mass of each beaker with water. A minimum of three trials is required.

## Calculations for Parts A – C

Before you can perform Part D of this experiment, you must complete the following calculations for Parts A – C. For each of the trials in a given Part, calculate the mass (g) and volume (mL) of water to the proper number of significant figures. The mass of water is the difference between the mass of the beaker with and without water. The volume of water can be calculated using the mass and density of the water at a specific temperature (see Appendix 11 for density values). *For the volume data only*, calculate the average, the range and the percent relative range for Parts A – C. Using the average volume of a given Part, calculate the error and % error of your measurements. Assume the known value for the volume of water is 10.00 mL in all cases. The percent error is a measure of accuracy, and the percent relative range provides a quick method to estimate precision. (Refer to Appendix 3 for more information about calculations.)

## Part D: Identity of an Unknown Pepsi™ Sample

The density of a solution depends on the amount of solute in the solution. In general there is a smooth relationship between the density and the weight percent of solute in the solution. For example, Diet Pepsi™ (containing no sugar), has a density of 0.994 g/mL at room temperature, while regular Pepsi™ (containing 13% sugar by mass) has a density of 1.040 g/mL. Thus, an accurate measurement of the density of an unknown Pepsi™ sample can be used to identify the type of Pepsi™ present.

Examine your results in Parts A – C and select what you think is the "best" method to measure the volume of a liquid. Make your choice of method by considering factors like the accuracy and precision, as shown in the determination of the volume of water. In the event two methods give comparable accuracy and precision, consider the ease or speed of use. Obtain a volume of the unknown Pepsi™ sample (40-50 mL) to perform this part and enter the number (if there is more than one sample available) in your notebook.

Using your chosen method (Part A, B, or C) determine the density of the unknown Pepsi™ using at least three samples. Whatever method you choose, make sure that you rinse out the glassware with unknown Pepsi™ to remove any residual water, as demonstrated by your instructor, before you make your measurements.

When finished, rinse the glassware thoroughly with deionized water before returning it to its place in lab.

Experiment 3

# What is the Concentration of Blue Dye?

## PURPOSE

To develop an experimental method for the determination of the concentration of blue dye are in an unknown solution.

## INTRODUCTION

Your instructor will give you a brief synopsis of this experiment during the laboratory lecture.

## PROCEDURE

**Special Equipment:** Spectrophotometer and cuvets

**Materials:** Stock blue dye solution (concentration 100 drops), unknown blue dye solution

**Laboratory Safety:**

- Be sure that your safety goggles are in place during all parts of this experiment.

There is not a set protocol for this experiment. You and your laboratory partner must devise a method to determine the concentration of blue dye in the unknown solution using materials/equipment available in the laboratory. In addition to the glassware available in your drawer, you will also have access to a spectrophotometer. Before coming to lab, you must read Appendix 12 which details the basic function of a spectrophotometer. Although you will work with a partner, you must keep an individual record of the experiment. The spectrophotometer should be set to 610 nm for all measurements. ALL observations you make during the experiment should be recorded in ink directly in your laboratory notebook.

## Experiment 4

# A Copper Cycle of Reactions

## PURPOSE

To recognize that change of state, change in color, formation of a precipitate, or the evolution of heat are associated with a chemical change; to study reactions of copper.

## INTRODUCTION

Chemical reactions involving metals often result in an observable change such as formation of a precipitate (an insoluble solid), evolution of gas, change in color, or change in temperature. As shown in the diagram below, you will begin and end this experiment with elemental copper (Cu) metal, but will transform the copper in a series of reactions into a variety of copper-containing compounds. According to the Law of Conservation of Matter, you should end up with the same amount of copper that you started with.

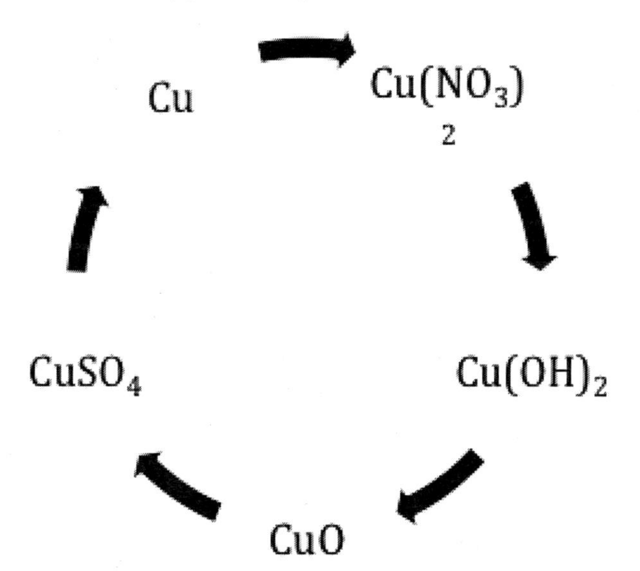


Your goal for this experiment is to make detailed observations during all parts of the experiment and interpret these in terms of the cycle given above. These should be recorded *clearly* and *completely* in your laboratory notebook in a *timely* fashion (i.e., as soon as you make them). Don't forget that if you are trying to detect *changes*, that means that you need to make observations both before and after you perform an operation. Your notebook must be well organized! The color of everything (solid, liquid, and gas) should be recorded, as should other physical appearances and changes including formation or disappearance of a solid, bubbling, and change in temperature. (In your descriptions, note that the word "clear" does not denote a color, but rather the absence of suspended solid material. It is synonymous with "transparent," not with "colorless." There is no contradiction in describing a solution as (e.g.) "clear and yellow.") The observational skills you are developing here will be put to use during the entire course.

## PROCEDURE

**Special Equipment:** Hotplate

**Materials:** Cu(s), Al(s), 6 M $HNO_3$, 6 M NaOH, 6 M $H_2SO_4$, 6 M HCl, acetone

**Laboratory Safety:**

- Be sure that your safety goggles are in place when working in the laboratory.
- Concentrated nitric, sulfuric, and hydrochloric acids ($HNO_3$, $H_2SO_4$, and HCl, respectively) and sodium hydroxide (NaOH) are corrosive and will rapidly react with skin and clothing. In particular, nitric acid will produce (non-permanent) yellow/orange spots when it contacts skin. Alert your instructor to any spills, and rinse any exposed skin or clothing with copious quantities of water.
- Acetone is flammable; be sure that no open flames are present when you are using it.
- All solutions containing copper are toxic.
- Be sure to dispose of all mixtures as directed below.

Record all observations you make during all parts of this experiment in your laboratory notebook. Copy the reaction cycle given above into your laboratory notebook. As you proceed through the steps below, add to the cycle by showing the chemical formulas and physical states of the materials added in each step, and by adding the descriptions of the product(s) formed in each step.

Weigh a clean, empty 100-mL beaker to the nearest (0.000 g). Add approximately 100 mg (0.100 g) of solid copper (Cu) (in the form of wire, foil, or granules) to the beaker and reweigh to the nearest mg. Record both masses and determine the mass of copper by difference.

**Reaction 1:** $Cu(s) + 4\ HNO_3(aq) \rightarrow Cu(NO_3)_2(aq) + 2\ NO_2(g) + 2\ H_2O(l)$

**Note:** The designations (*s*), (*l*), (*g*), and (*aq*) mean that the material is a pure solid, liquid, or gas, or an aqueous (water) solution, respectively.

Take the beaker and copper sample to the fume hood. One of the reaction products is a poisonous gas which you should avoid breathing and which should be vented through the fume hood.

**WORKING UNDER THE HOOD**, place the beaker on a white background (such as a sheet of notebook paper or a paper towel) to make it easier to observe the color of the reacting mixture. Add 6 M nitric acid ($HNO_3$) solution dropwise to the copper metal in the beaker. Start by adding 50 drops of the 6 M $HNO_3$ solution to the copper metal. (With the dropper bottles provided, approximately 25 drops equals one mL.) Swirl gently and cover the beaker with a watch glass or larger beaker. Record any changes (physical differences, color, temperature) that occur. When the reaction appears to slow down, add an additional 25 drops of $HNO_3$ solution. If necessary, continue adding $HNO_3$ until the copper metal is completely dissolved. Once this has occurred, remove the covering watch glass or beaker and allow the gaseous product to dissipate in the hood.

Once the gaseous product has dissipated, you can return to your lab bench with your beaker. The remainder of the experiment can be performed outside of a fume hood at your lab bench.

**Reaction 2:** $Cu(NO_3)_2(aq) + 2\ NaOH(aq) \rightarrow Cu(OH)_2(s) + 2\ NaNO_3(aq)$

Add 25 drops of 6 M sodium hydroxide (NaOH) solution to the beaker. Using a glass stirring rod, mix the contents thoroughly and record any changes. Repeat this process four more times. Add 25 mL of deionized water to the beaker, stir the mixture with the stirring rod, rinse any adhering solid from the stirring rod into the beaker with a stream of deionized water from a squirt bottle, and allow the solid to settle for 2-3 minutes.

**Reaction 3:** $Cu(OH)_2(s) \xrightarrow{\Delta} CuO(s) + H_2O(l)$

**Note:** the symbol "Δ" (capital Greek letter "delta") by the reaction arrow signifies that the reaction mixture was heated.

Place the beaker on a hotplate set to medium heat and set a watchglass on top as a cover. Allow the beaker to heat for 5-10 minutes, but do not allow it to boil. Swirl the beaker occasionally to prevent "bumping" of the mixture. If the solution starts to boil, temporarily remove the beaker from the hotplate, lower the hotplate setting, then return the beaker to the hotplate. Record any changes that occur. After no further changes appear to be occurring, remove the beaker from the hotplate and replace it with another beaker containing about 40 mL of deionized water.

Allow the solid to settle in the beaker, then decant as much as possible of the supernatant solution (that is, the solution above the solid) into an empty beaker. Try to retain as much of the solid as possible in the beaker. To the solid, add 40 mL of hot (not boiling) deionized water, swirl or stir the mixture, and allow the solid to settle for several minutes. Decant the supernatant solution into the beaker containing the first supernatant which you decanted, again trying to retain as much of the solid as possible in the beaker. The solution that was removed should be discarded into the "Heavy Metals Waste" container.

> **Note:** you can either discard this solution now, or you can continue to add material to your waste beaker and dispose of the wastes from various parts at one time later in the lab period.

**Reaction 4:** $CuO(s) + H_2SO_4(aq) \rightarrow CuSO_4(aq) + H_2O(l)$

Add approximately 50 drops of 6 M sulfuric acid ($H_2SO_4$) solution to the mixture in the beaker, swirl, and wait one minute. Record any changes observed. Add additional $H_2SO_4$ dropwise if needed until no further changes are observed and the reaction is complete.

**Reaction 5:** $3\ CuSO_4(aq) + 2\ Al(s) \rightarrow 3\ Cu(s) + Al_2(SO_4)_3(aq)$
$(2\ Al(s) + 6\ HCl(aq) \rightarrow 3\ H_2(g) + 2\ AlCl_3(aq))$

Obtain one piece of aluminum (Al) foil about 3.5 cm square, tear in half, crumple each piece loosely (do not fold), and add them to the beaker. Add 5 drops of 6 M hydrochloric acid (HCl) solution. Record any changes that occur. Gently swirl the beaker occasionally. If no signs of reaction are apparent, add a few more drops of HCl until the aluminum foil is completely dissolved. Use a glass stirring rod to break up any clumps of product, then rinse the stirring rod with deionized water.

**DRYING THE COPPER:** Decant the supernatant solution into a waste beaker transferring as little as possible of the solid into the waste beaker. Rinse the solid with 25 mL of deionized water and carefully decant the rinse into the same beaker. Rinse the solid a second time with 25 mL of deionized water and decant into the same beaker. Discard the solution into the "Heavy Metals Waste" container.

Rinse the solid with 5 mL of acetone ($C_3H_6O$) and decant into an empty beaker (not the same one you have been using to collect the previous waste solutions, since this waste will be placed into a different container). Rinse the solid a second time with 5 mL of acetone and decant into the same beaker. Place the beaker on a hotplate set to medium until the acetone has all evaporated and the solid is dry. (Don't forget to turn off the hotplate when you are finished.) Cool the beaker to room temperature on the lab bench (this typically takes about five minutes) before weighing it. Measure the mass of the beaker and solid product and determine the mass of product formed by difference. Discard the acetone rinses into the "Organic Waste" container. Discard the solid product into a laboratory trash can.

Experiment 5

# Synthesis and Limiting Reactant

## PURPOSE

To synthesize a compound containing nickel and to determine the percentage yield for the preparation using the limiting reactant concept.

## INTRODUCTION

An important application of the mole concept relates to the synthesis or preparation of chemical compounds. Two or more materials react with one another in a definite ratio by mass or by moles, and this relationship is expressed symbolically with a balanced chemical *equation*.

In this experiment, you will prepare a compound with the formula $Ni(C_2H_8N_2)_3Cl_2$ from nickel(II) chloride hexahydrate ($NiCl_2{\bullet}6H_2O$) and ethylenediamine ($C_2H_8N_2$) according to the balanced equation.

$$NiCl_2{\bullet}6H_2O + 3\ C_2H_8N_2 \rightarrow Ni(C_2H_8N_2)_3Cl_2 + 6\ H_2O \quad (1)$$

The name for the product, $Ni(C_2H_8N_2)_3Cl_2$, is trisethylenediaminenickel(II) chloride. It belongs to a class of compounds known as *coordination compounds*.

The reactant, $NiCl_2{\bullet}6H_2O$, (named nickel(II) chloride hexahydrate) belongs in a class of compounds known as *hydrates*. A *hydrate* is a compound that contains water molecules as an integral part of its structure. The "$\bullet 6H_2O$" indicates that there are 6 moles of water for every mole of $NiCl_2$. The formula weight (or molar mass) of this compound includes the mass of these 6 moles of water. Although it contains a large number of water molecules, the compound is a solid, as is true of almost all hydrates.

In any chemical reaction, if one of the reactants is totally used up, the reaction stops and no more products can be made. The above equation tells you that 1 mole of $NiCl_2{\bullet}6H_2O$ requires 3 moles of $C_2H_8N_2$ to react completely to form 1 mole of $Ni(C_2H_8N_2)_3Cl_2$ as a product. If you add more than three moles of $C_2H_8N_2$ per mole of $NiCl_2{\bullet}6H_2O$, the $C_2H_8N_2$ is present in excess and the $NiCl_2{\bullet}6H_2O$ acts as the *limiting reactant*. In other words, the $NiCl_2{\bullet}6H_2O$ limits the amount of product because it is totally consumed. On the other hand, if there are fewer than 3 moles of $C_2H_8N_2$ per mole of $NiCl_2{\bullet}6H_2O$, then the $C_2H_8N_2$ is the limiting reactant.

Once the identity of the limiting reactant has been determined, we can calculate the maximum amount of product that can be produced from the given amounts of starting materials. This amount is called the *theoretical yield* of the reaction. It is calculated by using the number of moles of the limiting reactant to compute moles of product that can be formed, based on the mole ratio shown in the balanced equation. From the maximum number of moles of product that can be formed, the maximum mass of product can also be calculated. (More details are given in the Calculations section below.)

When a reaction is carried out in the lab, the *actual yield* of the product isolated is normally less than the theoretical yield because of spills, impurities in the reactants, incomplete reactions, solubility of the product in the solvent used, and various other causes. A measure of the relative amount of product obtained, called the *percent yield*, is calculated from the relationship.

$$\text{Percent Yield} = \frac{\text{Actual Yield}}{\text{Theoretical Yield}} \times 100 \tag{2}$$

In this experiment, you will prepare $Ni(C_2H_8N_2)_3Cl_2$ by reaction (1) and determine the actual and percentage yields of the preparation.

## PROCEDURE

**Special Equipment:** Suction-filtration apparatus, filter paper, powder funnel, glass vial with cap

**Materials:** $NiCl_2{\bullet}6H_2O(s)$, 25.0 wt % ethylenediamine solution ($C_2H_8N_2(aq)$), acetone($l$)

**Laboratory Safety:**

- Be sure that your safety goggles are in place during all parts of this experiment.
- 25.0 wt % *ethylenediamine* solution is irritating to the skin; use with caution.
- *Acetone* is flammable; keep away from open flames.

### Preparation of Compound:

In a 100-mL or 150-mL beaker, accurately determine the mass of a sample of nickel(II) chloride hexahydrate ($NiCl_2{\bullet}6H_2O$) in the mass range 2.5–3.0 g, and record this mass *precisely* in your laboratory notebook. Note and record the appearance of the compound. If you accidentally spill the nickel compound on the balance or the table, please sweep it up with one of the brushes available in the balance room. Ask your instructor for help if necessary.

Using a glass rod for stirring, dissolve the sample completely in a minimum amount of deionized water. (Use no more than 6 mL of water, because the product is water-soluble and some will be lost if too much water is used.) Note and record the color of the solution. Obtain 8-11 mL of 25.0 wt % ethylenediamine solution in a graduated cylinder and record its precise volume in your notebook. With constant stirring with a glass stirring rod, slowly add it to the beaker containing the nickel(II) chloride hexahydrate solution. Note and record any changes that take place.

To the resulting solution, add 25 mL of acetone in 5-mL portions, and stir *well* after adding each portion. Continue to stir until the product begins to precipitate. Note and record the changes that occur. The product usually begins to precipitate during the addition of the acetone, but it is possible that you will have to stir for a few minutes, and possibly firmly scratch the inner walls of the beaker with the glass rod to initiate precipitation.

### Isolation of Compound:

Prepare an ice-water bath using approximately 2 parts ice to 1 part water. After precipitation has begun, place the beaker into the ice-water bath and allow it to stand undisturbed for a few minutes to cool. (Take care that the beaker does not tip over!) While the mixture is cooling, assemble the apparatus for suction filtration as described in Appendix 5. Note and record the appearance of the mixture before filtering.

Once the apparatus is assembled and the mixture is ready to be filtered, completely wet the filter paper, using as small an amount of de-ionized water as possible, then turn on the vacuum. Stir or swirl the slurry of product (the mixture of solid and liquid) and pour the mixture into the funnel. If (and only if) some of the solid gets around or through the filter paper at first, turn off the suction, remove the funnel from the flask, and pour the filtrate (the liquid that passed through the filter) back into the beaker and re-filter the sample. Try to transfer as much of the solid as possible to the funnel. If any solid remains in the beaker, add a few mL of acetone (not water) to the beaker, scrape the walls with a rubber policeman to loosen the solid, and add to the funnel. Allow the vacuum to pull through the solid for a minute or two to help evaporate the acetone and water.

### Purification of Compound:

You will need to wash the solid product to remove excess reactants and solvent. This is accomplished by turning off the suction, gently breaking up the solid with a metal spatula (be very careful not to tear or puncture the filter paper), and adding 5 mL of acetone to the solid. Be sure to contact as much of the solid as possible with this wash liquid. Then turn the vacuum back on to remove the wash liquid. (Do *NOT* re-filter regardless of whether or not your filtrate looks cloudy at this point.)

Repeat the *entire* washing process (vacuum off, add wash liquid, vacuum on) with two additional 5 mL portions of acetone. *Turn off the suction, place the funnel containing your product in a beaker for safe keeping and pour the contents of the filter flask into the marked waste container. Reassemble the apparatus, then turn the vacuum back on to pull air through the solid in the funnel for* ***at least*** *10-15 minutes (****the longer the better****) to help dry the product.* During this process, stir the solid gently, using a spatula to break up any lumps that might be present, and allow good air contact with the entire sample, but being careful not to tear or puncture the filter paper. The product should be a free-flowing powder. You should begin your calculations for the theoretical yield while you are waiting for the product to dry. The calculations should take no more than 30 minutes. By the time you finish, the product will be dry and you may proceed to the next step.

## Determination of Yield of Product:

Note and record the appearance (color and texture) of the product. Label a small vial with your name, section number and name of the product, then record the mass of the empty vial to ± 1 mg (means the same as 0.000g). You may choose to weigh the vial either with or without its cap. It is important only that you weigh the vial in the same condition (i.e., with or without the cap) both empty and with the dry product.

Carefully remove the funnel from the filter flask, and carefully transfer the product to the pre-weighed labeled vial using a powder funnel. Record the net mass of the product and submit the labeled vial to your instructor.

## Useful Calculation Information

The 25.0 wt % ethylenediamine solution had a density of 0.950 g/mL. A 25.0 wt % solution also contains 25 g of ethylenediamine (solute) per 100 g of solution. Mass of solution does NOT equal mass of solute!

First calculate the mass of the $C_2H_8N_2$ solution from its volume and density, then calculate the mass of the $C_2H_8N_2$ by remembering that

$$\text{wt \% solution} = \frac{\text{mass of solute}}{\text{mass of solution}} \times 100 \qquad (3)$$

From their masses and formula weights, you should then compute the number of moles of $NiCl_2 \cdot 6H_2O$ and of $C_2H_8N_2$ that were used. (Remember the formula weight of $NiCl_2 \cdot 6H_2O$ should include the mass of the waters of hydration.) From this point, there are a number of strategies which can be used to determine the limiting reactant and theoretical yield.

## Experiment 6

# Chemical Reactions

### PURPOSE

To become familiar with and classify types of chemical reactions by performing a series of procedures, recording observations, and relating observations to chemical equations.

### INTRODUCTION

Basic to the study of chemistry is the concept of a *chemical reaction*. In Experiment 3, you investigated a series of chemical reactions involving copper and its compounds. In this experiment, you will use those same ideas to investigate a wider range of reactions. In a chemical reaction, the composition and/or structure of some sample of matter undergoes a change. This change in the composition or structure is often signaled on some observable level by the change in one or more of the properties of the sample of matter. You will be using your sense of **smell** to detect odors; your sense of **sight** to detect bubbles of gas, the formation of a precipitate, or color changes; your sense of **hearing** to detect bubbling; and your sense of **touch** to detect temperature changes (but only on the *outside* of the reaction vessels!). It should also be noted that not every possible combination of reactants leads to an observable reaction. In many cases, two materials simply do not react with one another, and lead to the conclusion of "No Reaction".

From changes detected by our senses (or by our senses aided by an instrument), we infer the changes occurring at the atomic or molecular level. This leads us to some general ideas concerning the chemical reactivity of certain materials and the types of reactions they undergo. There are over 50,000,000 known chemical compounds, so the number of different possible chemical reactions is well over $10^{15}$. Of this huge number of possible reactions, it turns out that most can be classified into a very limited number of categories. In this experiment we will restrict ourselves to three of the categories: *precipitation*, *acid-base*, and *oxidation-reduction*.

#### Precipitation Reactions:

This type of reaction is often classified as a double replacement or metathesis reaction, generally involving two soluble, strong electrolytes. In this type of reaction, the cations and anions of the two reactants "switch partners." If one of the products is insoluble in water, it is called a precipitate, which is defined as an insoluble material formed from the mixing of aqueous solutions of two soluble materials.

Given the formulas of two ionic compounds, one can always write a chemical equation for this type of precipitation reaction. However, being able to write a chemical equation does not mean the reaction will actually occur. To predict whether or not a precipitation will take place, one must determine whether the products will be soluble or insoluble in water. This is done by referring to the solubility rules given in Appendix 8 or your textbook.

For example, the reaction of aqueous solutions of lead(II) acetate and sodium sulfate is a pre-cipitation-type reaction. One would predict the products to be formed by the cations ($Pb^{2+}$ and $Na^+$) "switching partners" to form lead(II) sulfate and sodium acetate:

$$Pb(CH_3CO_2)_2\ (aq) + Na_2SO_4\ (aq) \longrightarrow PbSO_4 + 2\ NaCH_3CO_2$$

Whether or not the reaction will actually take place depends on the solubility of $PbSO_4$ and $NaCH_3CO_2$. Examination of the solubility rules shows that $PbSO_4$ is insoluble and $NaCH_3CO_2$ is soluble, thus the equation can be completed with the predicted physical states of the products:

$$Pb(CH_3CO_2)_2(aq) + Na_2SO_4(aq) \longrightarrow PbSO_4(s) + 2\ NaCH_3CO_2(aq) \quad (1)$$

The reaction *should* occur since one of the products is insoluble. Indeed, if aqueous solutions of lead(II) acetate and sodium sulfate were to be mixed, the formation of a white precipitate would be observed. Refer to your textbook for further discussion and examples.

## Acid-Base Reactions:

Acids and bases can be defined in many ways, but, for our purposes we will define an acid as any species that can transfer a proton ($H^+$ ion) to another species, and a base is any species that can accept a proton from another species.

Most acids are readily recognized from the formula by the H that is written at the beginning of the formula, such as **H**Cl and **H**$NO_3$. One exception that you will often come across is acetic acid, an organic acid, which is often written as $CH_3CO_2$**H** (also written as **H**$C_2H_3O_2$). The H shown in boldface is called a *protic hydrogen*. It is the H being transferred in an acid-base reaction. (You will come across more of these organic acids in Chemistry 132 and 132L.)

Bases commonly contain the hydroxide ion ($OH^-$), such as NaOH and $Ca(OH)_2$. However the presence of the hydroxide ion is not necessary for a substance to act as a base. For example, $Na_2CO_3$, $KHCO_3$ and $NH_3$ can react as bases as well. You will be using compounds similar to these in this experiment.

An *acid-base* reaction is a reaction where a *hydrogen ion ($H^+$) is transferred*. The most common type of acid-base reaction is a double replacement reaction involving an acid as one of the reactants. The protic hydrogen of the acid is one of the cations involved in the double replacement. An example is the reaction of calcium hydroxide with hydrochloric acid. The $Ca^{2+}$ and $H^+$ ions "switch partners" to produce $CaCl_2$ and HOH (which is usually written as $H_2O$). The solubility rules show that $CaCl_2$ is soluble, thus, the reaction would be written as shown below.

$$Ca(OH)_2(aq) + 2\,HCl(aq) \longrightarrow CaCl_2(aq) + 2\,H_2O(l) \qquad (2)$$

$CaCl_2$ is classified as a salt, which is an ionic compound made up of a cation other than $H^+$ and an anion other than $OH^-$ or $O^{2-}$. Thus, in this type of double replacement, the products typically consist of water and a salt. Below is another example, involving acetic acid in place of hydrochloric acid, producing water and the salt, sodium acetate. (Note where the protic H is located in the formula of acetic acid.) The physical state of the salt is predicted according to the solubility rules.

$$NaOH(aq) + CH_3CO_2H(aq) \longrightarrow Na(CH_3CO_2)(aq) + H_2O(l) \qquad (3)$$

There are several common variations to these acid-base reactions with which you need to be familiar. If the product of the double replacement is carbonic acid ($H_2CO_3$), it should be replaced with $H_2O$ and $CO_2$. Carbonic acid is unstable in water and immediately decomposes to water and carbon dioxide.

$$H_2CO_3(aq) \longrightarrow H_2O(l) + CO_2(g) \qquad (4)$$

Thus, the reaction of hydrochloric acid and potassium carbonate is written as follows:

$$2\,HCl(aq) + K_2CO_3(aq) \longrightarrow 2\,KCl(aq) + H_2O(l) + CO_2(g) \qquad (5)$$

Note that $K_2CO_3$ is acting as a base, even though it does not contain $OH^-$. The $H^+$ is transferred from HCl to $CO_3^{2-}$, to initially form $H_2CO_3$, which then decomposes to water and carbon dioxide according to reaction 4.

## Oxidation-Reduction Reactions:

The third class of chemical reactions is the *oxidation-reduction* or *redox* reaction. *Oxidation* is defined as a process involving an increase in oxidation number (or loss of electrons); reduction, decrease in oxidation number (or gain of electrons). (Refer to your textbook for a discussion of oxidation numbers and other meanings of *oxidation* and *reduction*.) The criterion for a reaction to be classified as redox is there must be a *change in oxidation number* of one or more (usually two) elements in the reactants, due to a transfer of electrons.

We will consider a common group of redox reactions known as *single replacement* reactions, involving the reaction of an *element* with a *compound*. This type of reaction can occur when the element is a nonmetal (especially the halogen elements), but here we will be concerned only with reactions where the element is a metal. This elemental metal can replace either the metal cation or $H^+$ in the other reactant. The metallic element is converted from a neutral atom into its ionic form (Al to $Al^{3+}$ in equation 6) and Zn to $Zn^{2+}$ in equation 7), so is being oxidized. The element being replaced in the *compound* is converted from its ionic form to its neutral form ($Fe^{2+}$ to Fe in equation ) and $H^+$ to $H_2$ in equation 7), so is being reduced.

$$2\,Al(s) + 3\,Fe(NO_3)_2(aq) \longrightarrow 3\,Fe(s) + 2\,Al(NO_3)_3(aq) \qquad (6)$$

$$Zn(s) + 2\,HBr(aq) \longrightarrow ZnBr_2(aq) + H_2(g) \qquad (7)$$

As with precipitation reactions, one can *write* single replacement equations for any combination of an element with a compound. To predict whether or not the reaction will actually take place, one must examine the position of the elemental reactant with regard to the ionic reactant in the activity series of metals (see Appendix 9). The higher the element is found in the series, the more reactive it is. In general, this type of reaction will occur only if the *elemental* metal is high-er in the series (i.e., more reactive) than the metal in the compound (or the H in the case of an acid, as shown in equation 7). In equation 6, Al is more reactive than Fe, and in equation 7, Zn is more reactive than H, so both reactions are expected to take place. Refer to your textbook for more details and examples.

## PRE-LABORATORY ASSIGNMENT:

*Before you come to lab*, copy the table on the next page into your lab notebook and complete it. If necessary, make corrections in your lab notebook as this is discussed in the pre-lab discussion. When a formula is given, write the formulas of its ions (including the charges). When a pair of ions is given, write the formula of the ionic compound formed by these ions. In all cases, give a systematic name for the compound. Three of them have been completed (in italics) as examples. If you need help, review the relevant sections in your textbook and use Appendices 7 & 8 for reference.

| Formula of Compound | Formula of Cation (incl. the charge) | Formula of Anion (incl. the charge) | Name of Compound | Soluble in water? Yes or No |
|---|---|---|---|---|
| **$Na_2SO_4$** | *$Na^+$* | *$SO_4^{2-}$* | *sodium sulfate* | *Yes* |
| **KCl** | *$K^+$* | *$Cl^-$* | *potassium chloride* | *Yes* |
| **$Mg(NO_3)_2$** | | | | |
| **$K_3PO_4$** | | | | |
| **$Cr_2O_3$** | | | | |
| **$Na_2S$** | | | | |
| **$Na_2CO_3$** | | | | |
| **$Fe(NO_3)_3$** | | | | |
| **$Ca_3(PO_4)_2$** | | | | |
| **NaH** | | | | |
| *$Li_2CO_3$* | **$Li^+$** | **$CO_3^{2-}$** | *lithium carbonate* | *Yes* |
| | **$Ca^{2+}$** | **$Br^-$** | | |
| | **$Sn^{2+}$** | **$PO_4^{3-}$** | | |
| | **$Cu^{2+}$** | **$SO_4^{2-}$** | | |
| | **$Ni^{2+}$** | **$CH_3CO_2^-$** | | |
| | **$NH_4^+$** | **$SO_3^{2-}$** | | |
| | **$K^+$** | **$CrO_4^{2-}$** | | |
| | **$Mn^{2+}$** | **$CN^-$** | | |
| | **$Mg^{2+}$** | **$N^{3-}$** | | |

# PROCEDURE

**Special Equipment:** Test tube holder

**Materials:**

Group A: 0.1 M $Na_3PO_4$, 0.1 M NaI, 0.1 M $Na_2CrO_4$

Group B: 0.1 M $AgNO_3$, 0.1 M $Cr(NO_3)_3$, 0.1 M $Pb(NO_3)_2$

Group C: 1 M $AgNO_3$, 6 M HCl, 6 M KOH, 1 M $HNO_3$, 0.1 M $Al_2(SO_4)_3$, 1 M $Ca(CH_3CO_2)_2$, 0.1 M $Cr_2(SO_4)_3$, 0.1 M $CuCl_2$, Cu(s), Mg(s), Al(s), $NiCO_3$(s)

**Laboratory Safety:**

- Be sure that your safety goggles are in place during all parts of this procedure.
- Be aware that many routine laboratory chemicals are poisonous, particularly $Na_2CrO_4$, $AgNO_3$, $Cr(NO_3)_3$, and $Pb(NO_3)_2$.
- Acids and bases can cause burns and other types of eye and skin damage. *Hydrochloric acid*, *nitric acid* and *potassium hydroxide* are of particular concern.
- *Silver nitrate* will discolor your skin; avoid contact with the solution.
- Dispose of all of your solutions, precipitates, and excess reagents in the “Heavy Metals Waste” container.

## Part A: Precipitation Reactions

You will be working with aqueous solutions of six ionic compounds. Group A consists of the three compounds that have the $Na^+$ cation, and Group B consists of three with the $NO_3^-$ anion. You are to combine *one solution from Group A* with *one solution from Group B* following the procedure described below. Devise a method to ensure that you perform ALL the possible combinations. Before you begin, prepare a data table (as shown below) in your notebook to record the formulas and appearance of reactants and resultant mixture for each reaction. Be sure to leave ample space to record the “Appearance of Precipitate” so that you can expand on your observations later, if necessary.

| Formula of Reactants | Formula of Ions in Each Reactant | Appearance of Reactants | Appearance of Precipitate | Formula of ppt |
|---|---|---|---|---|
| A. $Na_2CO_3$ | $Na^+$, $CO_3^{2-}$ | colorless | ➢ Soln turned cloudy, pale green<br>➢ After 10 mins, green ppt settled on bottom | $NiCO_3$ |
| B. $Ni(NO_3)_2$ | $Ni^{2+}$, $NO_3^-$ | green | | |

Clean sufficient small test tubes to perform one test in each, rinse them with deionized water, and shake out the excess water. (The tubes do NOT need to be completely dry.) Obtain one aqueous solution from Group A and one from Group B. Copy the formulas and record the appearance of each reactant. (In recording the color, note that the word “clear” does not designate a color—it means that the material is a true solution and contains no suspended, insoluble material. “Clear” is the opposite of “cloudy.” A solution that does not have a color is said to be “colorless.”)

Add 10 drops of each to one of your test tubes. Thoroughly mix the contents of the tube by holding the tube firmly near the top with one hand, and then flick the side of the tube near the bottom with the other hand. Your instructor will demonstrate this technique. Do not shake it by covering the top of the test tube with your thumb and shaking the tube! Remember that if you can see definite layers in a mixture, it’s *not* thoroughly mixed.

After mixing, if the contents no longer look *clear* but have turned *cloudy* or *opaque*, it means that a precipitate has formed. If so, record its appearance and write the formula of the precipitate. Note: Not all mixtures will form precipitates. A change in color indicates that some type of chemical reaction has taken place, but does not mean a precipitate has formed.

After about 10 minutes, re-examine each of the test tubes. Add to your observations in your data table, if necessary. For example, by comparison of all the precipitates, you might decide to describe a particular precipitate as “bright yellow” or “pale yellow” instead of simply “yellow.”

## Part B: Additional Chemical Reactions

Thoroughly clean seven small test tubes, rinse them well with deionized water, and allow them to drain. It is not necessary to dry the tubes thoroughly unless doing so is specifically called for in the procedure. If you do need to dry a test tube, you may dry the tube by rinsing with a few milliliters of acetone. Acetone should be disposed of in the "Organic Waste" container.

*Before* combining the materials specified in Group C, be sure to record in your notebook complete, careful observations of the *initial properties* of each material. This would include the physical state, color, odor, and any other distinguishing characteristics you want to record. You need to have a good idea of the initial appearance if you want to determine changes in appearance.

In this experiment, it is not necessary to measure the amounts of materials with extreme care, but the amounts you use should be reasonably close to those specified. Liquids can be measured by counting drops (typically, about 25 drops ≈ 1 mL) and the masses of solids can be estimated. (The amount of a typical solid that just fills the rounded portion of the bottom of a small test tube, about the volume of a pea, is about 0.3 to 0.5 g.)

In checking for odors, the correct procedure is to hold the tube about 15 cm (6 inches) in front of your face, with the opening slightly below nose level, and fan the vapors toward your face. If no odor is detectable, slowly bring the tube closer to detect any less- obvious odors. Also be sure to use your sense of touch (but only on the *outside* of the test tube!) to check the temperature of the mixture before and after mixing, since a temperature change can indicate that a reaction has taken place. Clearly and completely record "before" and "after" observations in your notebook.

Re-examine each of the test tubes after the mixture has been left quietly standing for about 10 minutes. Remember that you may verify your observations for any reaction by repeating it.

## Reactions for Part B

1. Place 1.0 mL of 1 M silver nitrate in a test tube. Add a piece of copper wire or foil and record the results.

2. Place 1.0 mL of 6 M hydrochloric acid in a test tube. Add 1.0 mL of 6 M potassium hydroxide, mix well, and record the results.

3. Place 2.0 mL of 6 M hydrochloric acid in a test tube. Add one or two 1-cm pieces of magnesium ribbon and record the results.

4. Dry a test tube as directed above. Place about 0.1–0.2 g of solid nickel (II) carbonate into the tube. Add 2.0 mL of 1 M nitric acid dropwise, mixing well after each drop, and record the results.

### Reactions for Part B Cont.

5. Place 1.0 mL of 0.1 M aluminum sulfate in a test tube. Add a piece of copper wire or foil and record the results.

6. In a separate test tube, place 1.0 mL of 0.1 M copper (II) chloride and add a small piece of aluminum wire or foil and record the results.

7. Place 1.0 mL of 1 M calcium acetate ($Ca(CH_3CO_2)_2$) in a test tube. Add 1.0 mL of 6 M hydrochloric acid, mix well, and record the results.

# Experiment 7

# Solution Preparation

*This experiment was adapted from Eslek, Z.; Tuplar, A. Solution Preparation and Conductivity Measurements: An Experiment for Introductory Chemistry. *J. Chem. Ed.* **2013**, *90*, 1665.

## PURPOSE

To learn how to prepare solutions of known concentration (molarity) starting from a solid, and also by dilution of a stock solution. To measure the conductivity values of the prepared solutions as an evaluation of the solution preparation technique.

## INTRODUCTION

A solution is a homogeneous mixture created by dissolving one or more solutes in a solvent. Solutions with accurately known concentrations are called standard (stock) solutions. These solutions can be prepared by dissolving the desired amount of solute into a volumetric flask of a specific volume.

Stock solutions are frequently diluted to solutions of lesser concentration for experimental work. Dilution is the addition of more solvent to produce a solution of reduced concentration. A diluted solution is usually prepared from a small volume of a more concentrated stock solution according to Equation 1,

$$C_1V_1 = C_2V_2 \qquad (1)$$

where $C_1$ and $V_1$ refer to the concentration and volume of the stock solution used to prepare a dilution. Likewise, $C_2$ and $V_2$ refer to the concentration and final volume of the diluted solution. The moles present in the small volume of the stock solution (before dilution, 1), are equal to the moles present in the diluted solution created (after dilution, 2) as seen in Equation 2.

$$(\text{moles of solute})_1 = (\text{moles of solute})_2 \qquad (2)$$

In this experiment, sodium sulfate solutions of known concentrations will be prepared, and the conductivity values of the solutions will be measured with a LabQuest2 conductivity probe. In principle, a conductivity meter works by applying a voltage between two electrodes and measuring the current that flows between them. Cations move toward the negatively charged electrode, and anions move toward the positively charged electrode. Movement of these charged species results in an electrical current. The conductivity of a solution is proportional to the number of ions per unit volume, and therefore correlates to the concentration of an electrolyte-containing solution.

Given the relationship between measured conductivity and the molar concentration of a given species, a plot of conductivity vs. concentration will have a linear correlation. In this experiment, conductivity values will be measured in micro siemens per centimeter (μ/cm).

## PRE-LABORATORY ASSIGNMENT

Before you come to lab, clearly write out in your notebook all calculations necessary for the preparation of the three solutions listed in Part A. If necessary, corrections to these calculations can be made once in class. Be sure to pay close attention to units!

## PROCEDURE

**Special Equipment:** volumetric flasks (100 mL, 50 mL and 25 mL), 10-mL volumetric pipet, pipet pump, LabQuest2 kit with conductivity probe, power cable and 30 mL beaker

**Materials:** sodium sulfate, 0.10 M acetic acid ($CH_3COOH$) solution, 0.10 M sodium chloride solution

**Laboratory Safety:**

- Be sure that your safety goggles are in place during all parts of this procedure.
- Dispose of excess acid and acid solutions in the "Acid-Base Waste" container provided in the laboratory.
- All sodium sulfate and sodium chloride solutions may go down the sink.

### Part A: Preparation of Sodium Sulfate Stock Solutions

You will prepare three stock solutions starting from solid sodium sulfate following the procedure outlined below. Prepare the following stock solutions:

Solution #1: Make 100 mL of a 5.0 g/L sodium sulfate solution
Solution #2: Make 100 mL of a 0.025 M sodium sulfate solution
Solution #3: Make 100 mL of a 2.0 mM sodium sulfate solution

Calculate the mass (g) of sodium sulfate needed to prepare 100.0 mL of each solution. Once prepared, each solution will be stored in a tightly capped, 125 mL plastic bottle. Be sure to clearly label each bottle with the following information:

- Concentration (actual concentration based on mass of sodium sulfate used)
- Units
- Chemical species
- Student Initials
- Date prepared

## Preparing a Solution from a Solid (Figure 1):

1. Obtain a clean, dry 50 mL beaker, place it on the balance, and tare the balance so that it reads zeros for all values. Remove the beaker, and then using a spatula, add a small scoop of sodium sulfate to the beaker and then reweigh the beaker + solid. Repeat this process until the calculated mass of sodium sulfate is reached. It is not necessary to obtain the *exact* amount calculated, but for these solutions you should be within +/- 0.01 g of the calculated quantity. Be sure to record exactly how much solid was obtained.

   * Overshoot the Mass: If too much sodium sulfate is added to the beaker, DO NOT put it back in the original container. Instead, dispose of the sodium sulfate in the trash, clean the beaker and start again.

   ** Clean-up Balance Area: If you accidentally spill sodium sulfate on the balance or the table, please sweep it up with one of the brushes available in the balance room. Ask your instructor for help if necessary.

2. Add ~30 mL of deionized water to the beaker and gently swirl the beaker until all of the solid is dissolved (this can take a few minutes).

3. Fill a water bottle with clean DI water. Once the salt is dissolved, carefully pour all of the solution from the beaker into a clean, labeled 100.0 mL volumetric flask. Using the water bottle, carefully rinse the beaker with 4-5 mL of water, and add the rinse water to the volumetric flask. Repeat the rinsing step two more times to ensure quantitative transfer of the sodium sulfate from the beaker to the flask. If drops of water are lost during the transfer process, you will need to remake the solution.

4. Fill the volumetric flask with DI water until the bottom of the meniscus is level with the etched mark on the neck of the flask. Adding the last few milliliters using a Pastuer pipet allows for finer control. Stopper or cap the flask, and invert it several times to mix the solution thoroughly, each time allowing the air bubble in the flask to rise completely to the top. If too much DI water is added, causing the solution level to rise above the etched mark on the flask, you will need to remake the solution.

5. Calculate the *actual* concentration of the solution based on the *actual* mass of sodium sulfate used. Store the prepared solution in a clean 125-mL plastic bottle that is clearly labeled.

78.04 + 64

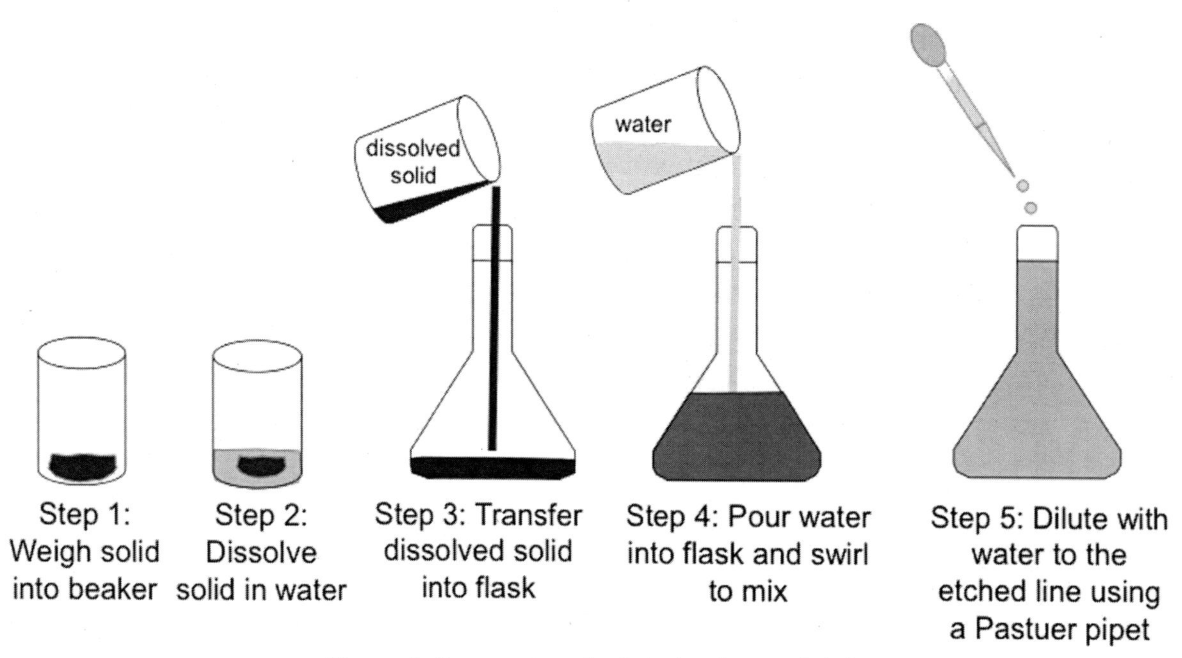


**Figure 1:** Preparation of a Solution from a Solid

## Evaluating the Stock Solutions:

Re-calculate the *actual* concentration of each stock solution in units of molarity (M). If time allows, measure the conductivity of the three prepared stock solutions of sodium sulfate following to the instructions in Part C below. Evaluate the graphed conductivity data for the three stock solutions and confirm that the correlation value is at least 0.998 or better. If the correlation value is *too low*, consult your instructor as you may need to re-make solutions.

## Part B: Preparation of Sodium Sulfate Dilutions

An X-fold dilution divides the concentration of a solution by a factor of X. For example:

A 5-fold dilution of a 0.25 M stock solution would result in a 0.050 M final solution.

A 2.5-fold dilution of a 0.25 M stock solution would result in a 0.10 M final solution.

Similarly, an X-fold dilution relates to the volume of stock solution needed to prepare a given quantity of a dilution. For example: You need to prepare 100 mL of a 5-fold dilution from a stock solution.

100 mL = total volume of diluted solution to be made

5-fold = dilution factor

To determine the volume of stock solution to use, divide 100 mL by 5 and you find that you need 20 mL of the stock solution to make the dilution. If the stock concentration in this example were 0.25 M, then the final concentration of the dilution would be found by using equation 1:

$$(0.25\text{ M}) * (20\text{ mL}) = (\text{dilution M}) * (100\text{ mL})$$

$$\text{Dilution concentration} = 0.050\text{ M}$$

You will use the stock sodium sulfate solutions from Part A to prepare three dilutions according to the procedure outlined below. Prepare the following three dilutions:

Solution #4: Make 25 mL of a 2.5-fold dilution of the 5.0 g/L sodium sulfate solution
Solution #5: Make 25 mL of a 2.5-fold dilution of the 0.025 M sodium sulfate solution
Solution #6: Make 50 mL of a 5-fold dilution of the 2.0 mM sodium sulfate solution

Each dilution will require 10 mL of the stock solution, delivered into the appropriately sized volumetric flask. Calculate the *actual* concentration of each of the final dilutions based on the information above and the *actual* concentrations of the stock solutions prepared in Part A. Be sure to clearly label each solution with the following information:

- Concentration
- Units
- Chemical species
- Student Initials
- Date prepared

### Preparing Dilutions (Figure 2):

1. Obtain a clean, dry small beaker (no larger than 50 mL), and pour approximately10 mL of the stock sodium sulfate solution into the beaker (clearly label the beaker). Rinse a 10-mL volumetric pipet with a few milliliters of the stock solution, and then discard the rinse into a waste beaker. Repeat the rinse process 2-3 times.

2 Refill the small beaker with approximately 15 mL fresh stock solution, and pipet 10 mL of the stock solution into a clean volumetric flask (flask size equals final volume of solution being prepared).

3. Fill the volumetric flask with DI water until the bottom of the meniscus is level with the etched mark on the neck of the flask. Stopper or cap the flask, and invert it several times to mix the solution thoroughly, each time allowing the air bubble in the flask to rise completely to the top. If too much DI water is added, causing the solution level to rise above the etched mark on the flask, you will need to remake the solution.

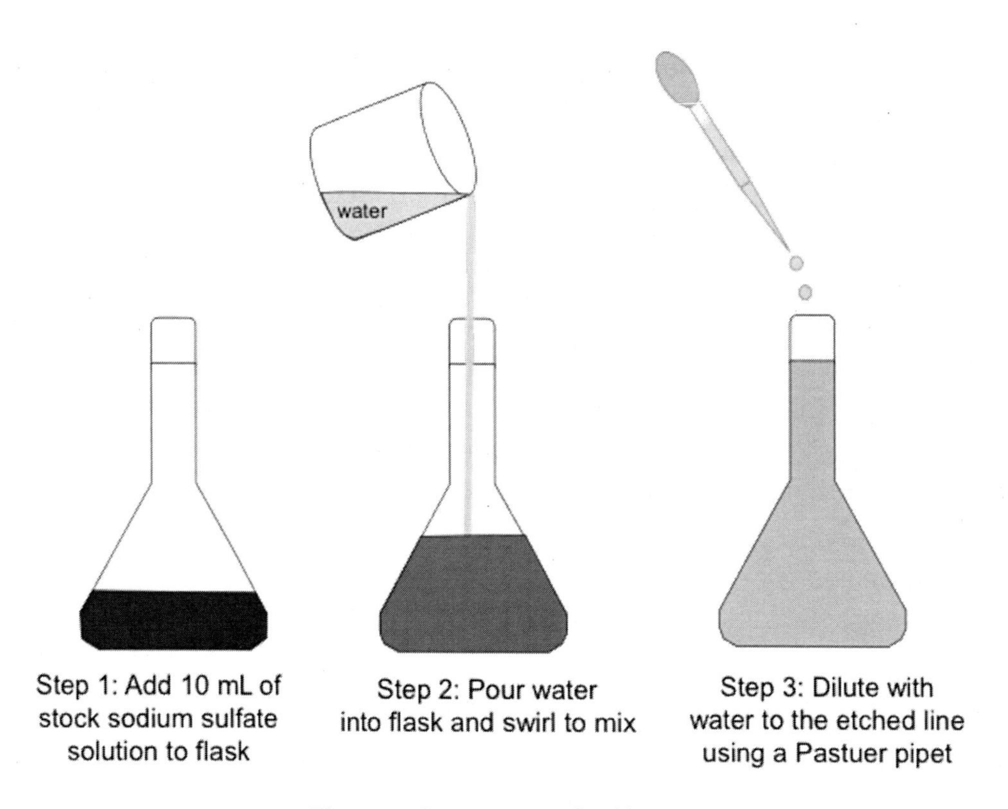


**Figure 2:** Preparation of a dilution

## Part C: Conductivity Measurements

**Sodium Sulfate Conductivity:** Connect the conductivity probe to the Channel 1 port (**CH1**) on the left edge of the LabQuest2. Make sure the toggle switch on the conductivity probe is set to the range 0-20,000 S. Attach the power cord to the **AC adapter port** on the right edge of the LabQuest2, and then plug it into the power source. Switch on the LabQuest2 by pressing the power button on the top left edge of the device. Once the LabQuest2 has been turned on you should see a screen similar to Figure 3. You now need to change the settings. Tap the **Mode** window on the right side of the screen and you should see a screen similar to Figure 4. Tap the pull-down menu and select **Events with Entry** (Figure 5).

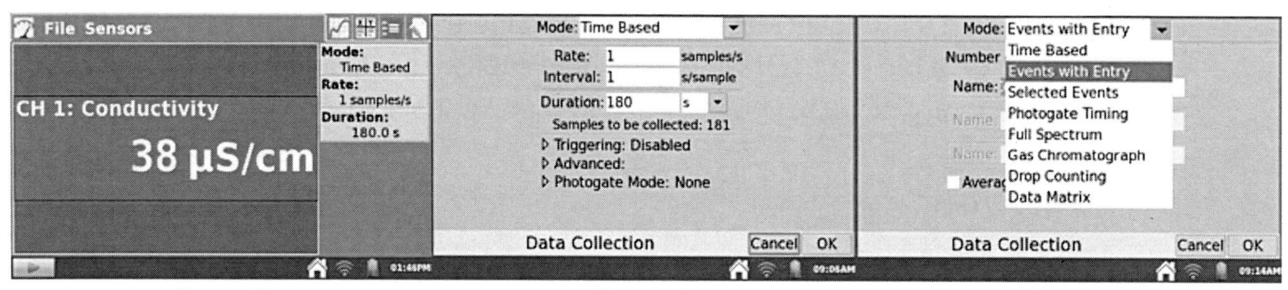


Figure 3 Figure 4 Figure 5

On the next screen, Figure 6, tap the **Name** space and using the letter keyboard, type in Concentration and tap **Done**. Tap the **Units** space and type M (for molarity) and tap **Done**. Tap **OK** and you should see a screen similar to Figure 7. To start collecting data, tap the start button (**green triangle**) in the lower left corner of the screen. You should now see a screen similar to Figure 8. Note that the real-time conductivity value is displayed in the grey box to the right of the screen.

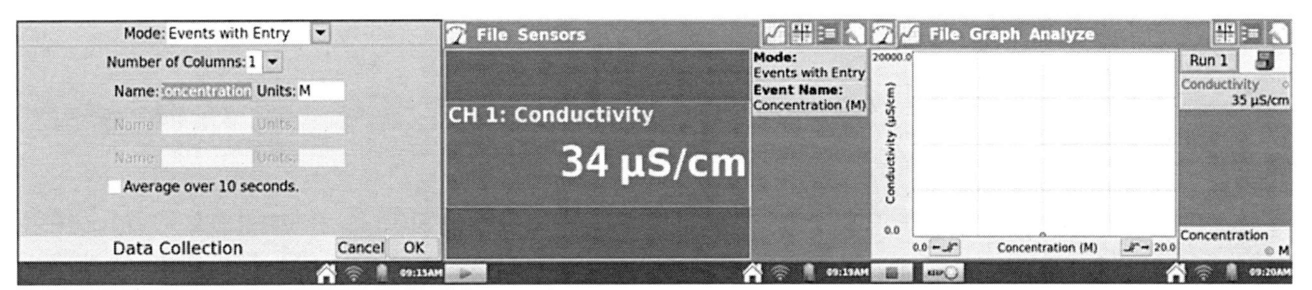


Figure 6 Figure 7 Figure 8

You will record conductivity measurements of the six sodium sulfate solutions that you prepared. Starting with your lowest concentration solution, pour approximately 25 mL into the clean, dry 30 mL beaker that was with the conductivity probe. Place the probe into the solution, making sure that the entire metal electrode is submerged. Swirl the probe in the solution to remove any trapped bubbles. When you are ready to record the conductivity value, tap **KEEP** and you should see a screen similar to Figure 9. Type in the solution concentration, and then tap **OK**. This should bring you to a screen similar to Figure 10. Between each solution measurement, empty, clean and dry the 30 mL beaker. Also, using a water bottle, rinse the conductivity probe with deionized water and gently dry the probe using a Kimwipe.

Repeat this process for all of your sodium sulfate solutions. After you have entered the last concentration, tap the stop button (**red square**) at the lower left hand corner of the screen. You should have a graph showing all the data points.

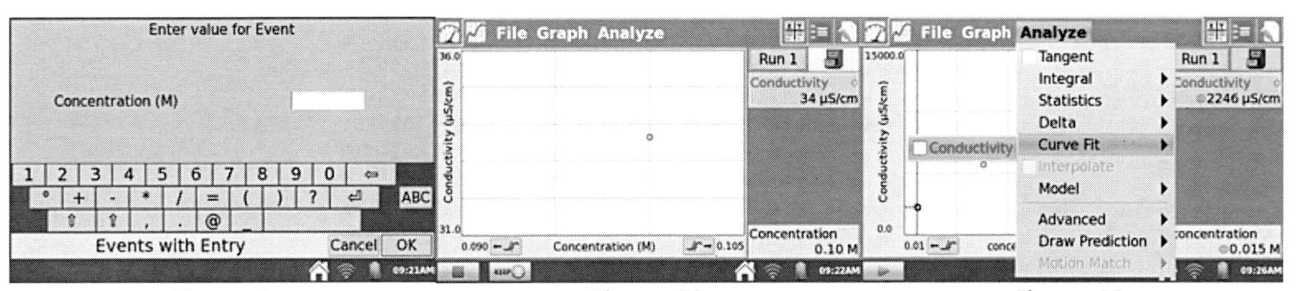


Figure 9 Figure 10 Figure 11

To analyze the data, tap **Analyze**, select **Curve Fit** and then tap the **Conductivity** tab as shown in Figure 11. Tap **Choose Fit** on the pull-down menu as seen in Figure 12, select **Linear** and then tap **OK**. You should now see a screen similar to Figure 13.

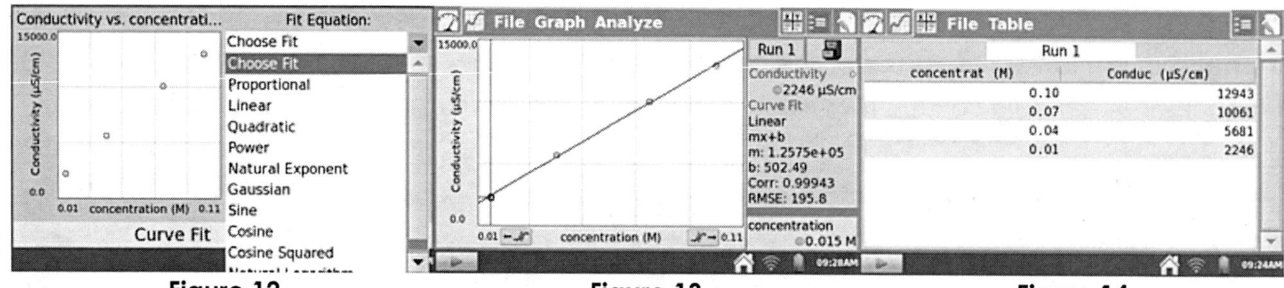


Figure 12 Figure 13 Figure 14

Evaluate the graphed data and confirm that the correlation value is at least 0.998 or better. If the correlation value is *too low*, consult your instructor as you may need to re-make solutions. If the correlation value is acceptable record it in your notebook, and tap the **Table** icon, third from the right on the Toolbar, to obtain a screen similar to Figure 14. Record the conductivity values (including units!) obtained in your notebook. Then, tap the **Meter** icon (first from the left) to return to the active measurement screen and proceed to the next section.

**Sodium Chloride and Acetic Acid Conductivity:** There are two additional solutions for which you will need to obtain conductivity values: a 0.1 M sodium chloride solution and a 0.1 M acetic acid solution. Both of these solutions are available in lab. For each solution, obtain approximately 25 mL in the clean, dry 30 mL beaker. Place the conductivity probe into the solution, making sure that the entire metal electrode is submerged. Swirl the probe in the solution to remove any trapped bubbles. Record the conductivity value observed for each solution in your notebook (these values do not need to be saved on the LabQuest2 device). Be sure to rinse and dry the conductivity probe before each measurement.

Once all conductivity values have been obtained, rinse all glassware with de-ionized water and return it to its appropriate locations. Any acetic acid should be placed in the "Acid-Base Waste" container in the lab. Solutions of sodium sulfate and sodium chloride may be rinsed down the sink.

Experiment 8

# Acid-Base Titration

## PURPOSE

To use a standard solution of sodium hydroxide in the volumetric analysis of an acid-containing sample. The ultimate goal is to determine the molarity and mass percent of acetic acid in an unknown vinegar sample.

## INTRODUCTION

An important application of stoichiometry in the laboratory is the analysis of materials to determine their composition. When the measurements are primarily of the volumes of solutions, the procedure is called a *volumetric* analysis. In this experiment, the techniques of volumetric analysis are discussed and applied to the analysis of an acid-containing sample.

In a typical volumetric analysis (frequently called a *titration*), the sample to be analyzed is placed in a flask or other container and then treated to prepare it for analysis. Once the sample is prepared, a measured volume of a solution of known concentration (the *titrant*) is added to cause some reaction of known stoichiometry to go *exactly* to completion. The moles of titrant are calculated from its volume and concentration, and then the moles of unknown material are calculated based on the stoichiometry of the reaction. A solution whose concentration is known precisely and which is used to determine the composition of an unknown sample is called a *standard solution*.

For a reaction to be usable in a volumetric analysis, it must have certain properties, including well defined stoichiometry, no side reactions, fast reaction rate, no significant interferences (that is, the material being analyzed should be the only material that reacts with the titrant), and some means of detecting the equivalence point of the titration. The *equivalence point* of a titration is the point at which the quantities of reactants are in the exact stoichiometrically-correct ratio for the reaction to go to completion. The *endpoint* of a titration is the point at which the titration is complete, as indicated by some sort of chemical indicator.

An *indicator* is a substance that changes color according to its chemical form. An *acid-base indicator* is a substance that itself has both acid and base forms, and each of which is a different color. In this titration, the development of pink color by the indicator *phenolphthalein*, as it goes from its acidic to basic forms, indicates the endpoint of the titration. The *endpoint* and the *equivalence point* are usually slightly different, as a small excess of titrant is necessary to get the indicator to change color. Care must be taken in selecting a chemical indicator so that the endpoint occurs as closely as possible to the equivalence point.

Many types of reactions can be used for titrations; precipitation, redox, and acid base reactions can all be successfully used under the proper conditions. In this experiment, you will use an acid-base reaction to analyze a sample containing an acid. As discussed in Experiment 5, an acid base or neutralization reaction involves the transfer of a proton ($H^+$ ion) from the acid to the base. The net ionic equation of the reaction of a strong acid and a strong base is the reaction of hydrogen ions with hydroxide ions:

$$H^+(aq) + OH^-(aq) \longrightarrow H_2O(l) \quad (1)$$

There are two parts to this experiment. In Part A you will learn to read a buret accurately, and in Part B you will use a buret filled with a standard NaOH solution to determine the concentration of acetic acid in an unknown sample of vinegar.

## PROCEDURE

**Special Equipment:** 50-mL buret, buret clamp, ring stand, racks of sealed buret barrels, 5-mL volumetric pipet

**Materials:** standard ~0.1 M NaOH, phenolphthalein solution, vinegar sample

**Laboratory Safety:**

- Be sure that your safety goggles are in place during all parts of this procedure.
- *Sodium hydroxide* solutions are corrosive to the eyes and can cause burns on skin if not flushed off with water; they also damage clothing.
- Dispose of excess acid and base (including all rinsings of your apparatus) in the "Acid-Base Waste" container provided in the laboratory.

*Before starting the experiment*, you should read the *entire procedure* and determine the kind of data you will be recording (e.g. concentration of standard NaOH, final buret reading (mL), initial buret reading (mL), volume of NaOH delivered (mL), description of end point, etc.). Plan and prepare a data table for all the trials so that you can enter your data in a neat and organized manner to make it easier to retrieve information when you need it.

### Part A. Reading Sealed Buret Barrels

Burets are used to deliver variable, precise volumes of liquids. The buret is a calibrated tube that can be filled with a liquid and is equipped with a valve/stopcock to control flow from the tip of the buret. While burets are available in many sizes, ranging from 1 to 100 mL, the 50-mL buret is most common in the General Chemistry laboratory. This buret is graduated in 0.1-mL increments. Thus, this buret can be read to ± 0.01 mL by estimating (interpolating) between the 0.1-mL calibrations clearly marked by lines on the buret. The volume delivered from the buret is calculated by the *difference* between the initial and final readings.

In this section your instructor will give you instructions on the use and accurate reading of the 50-mL buret. These are also given in Section E of Appendix 5. You will be able to test your buret-reading skills by comparing your buret readings with known buret readings.

Before you begin, you should prepare a data table in your laboratory notebook for the buret reading exercise as follows:

| *Buret-Pair Letter* | | |
|---|---|---|
| *Water Level in Buret #1 (mL)* | | |
| *Water Level in Buret #2 (mL)* | | |
| *Difference in Water Levels (mL)* | | |
| *Known Difference in Water Levels (mL)* | | |
| *Error in Difference in Water Levels (mL)* | | |
| *% Error* | | |

Sample Table Only. Record data directly in your lab notebook.

You will find several pairs of sealed buret barrels set up around the laboratory. One member of the pair represents the initial volume of the buret, while the other represents the final volume. Each pair is labeled with a letter. In the first three rows in your data table, record the identification letter and the water level inside for each of the paired, sealed buret barrels in one pair. Calculate the difference between these two readings and report the difference in the fourth row of your data table.

After you have completed this procedure for *two* pairs of buret barrels, obtain the known difference for each from your instructor and place this in the fifth row. Complete the calculations for error and % error. If your percent error is greater than 0.2% for either pair, ask your instructor for help with reading the burets, and then repeat the process for a third pair of buret barrels before proceeding to Part B.

### Part B: Determination of the Molarity and Percent Acetic Acid in Vinegar

The common kitchen "chemical" vinegar is an aqueous solution containing 4 to 6% by mass acetic acid, plus other coloring and flavoring agents. It has a density of 1.008 g/mL. Since the only acidic component in the mixture is acetic acid, you can easily determine the amount of this material by titrating with standard NaOH. The reaction (in molecular form) is

$$CH_3CO_2H(aq) + NaOH(aq) \longrightarrow H_2O(l) + NaCH_3CO_2(aq) \quad (2)$$

It is important to be able to distinguish between the terms "acetic acid" and "vinegar." "Acetic acid" refers specifically to the compound $CH_3CO_2H$ (sometimes written as $CH_3COOH$

or $HC_2H_3O_2$), whereas "vinegar" refers to a solution of acetic acid dissolved in water (i.e., $CH_3CO_2H(aq)$). Acetic acid is the solute, and vinegar is the homogeneous mixture of acetic acid and water. Care should be taken not to use the two terms interchangeably.

**Cleaning & Filling the Buret:** Clean a 50 mL buret and rinse it well with water, including the tip. Obtain approximately 200 mL of the ~0.1 M standard NaOH solution (be sure to record the exact concentration in your notebook) in a clean and dry beaker. Be sure to label this beaker and keep it covered with a clean and dry watch glass at all times. Close the stopcock and add to the buret about 5 mL of the ~0.1 M NaOH solution. Tilt the buret and rotate it so that this solution rinses the entire inner surface. Discharge the rinse through the tip into a waste beaker (**not** back into your beaker!). Repeat the rinsing process three times with fresh 5-mL samples of your NaOH solution. ***Do not use excessive amounts of your NaOH solution in the rinses***. For more information on how to use a buret properly, consult Appendix 5, Part E.

After rinsing three times, fill the buret to above the zero mark with the NaOH solution. If you use a funnel to do this, be sure to *remove it before making your initial reading*. Then drain the excess solution into a waste beaker (NOT back into the original beaker of NaOH solution) to remove any air bubbles in the tip, and to bring the level of solution down into the calibrated portion of the buret between the 0.0 and 1.0-mL marks.

**Preparing the Vinegar Solution for Titration:** Obtain a sample of vinegar from your instructor and *record its identification number* in your lab notebook. Clean four 250-mL Erlenmeyer flasks and rinse them with de-ionized water (the flasks do not need to be dry). Pour about 10 mL of the unknown vinegar sample into a clean, dry 50-mL beaker. Using this vinegar sample, rinse the 5-mL pipet *three times* with small portions (~2-3 mL) each time. Discard each rinse into your waste beaker. Finally, accurately pipet 5.00 mL of the vinegar sample into each flask. To each flask, add 3-5 drops of phenolphthalein indicator solution and 20–30 mL of de-ionized water.

## TITRATION PROCEDURE:

**Before you begin:** Empty a plastic wash bottle, then rinse and refill it with de-ionized water. It is good practice to have a uniform white background under the flask during the titration to make it easier to see the color change described below. This is easily done by placing a piece of white paper under the flask.

**Color change with phenolphthalein:** You will see a pink coloration where the NaOH solution hits the vinegar solution in the flask; and this coloration will disappear when the flask is swirled to mix the solution. When the pink color takes longer and longer to disappear, you are nearing the end point. Slow the rate of addition of NaOH and allow more time for mixing the two solutions. When nearing the endpoint, you should rinse the walls of the flask with a stream of de-ionized water from the plastic wash bottle, then swirl well to mix the solution. The end point occurs when the *entire* solution turns *pale* pink and remains pale pink (the paler, the better) for at least 20 seconds **after mixing well**.

**Scout titration:** Record the initial volume of NaOH in the buret. Then, open the stopcock and add the sodium hydroxide solution quickly to one flask of vinegar sample, while swirling the flask. When the pale pink endpoint is reached, close the stopcock and record the final volume of the NaOH solution in the buret. Calculate the total volume of NaOH added to the flask. This value should be used as an *estimate* of how much NaOH solution to add for each of the remaining titrations. Your instructor will demonstrate this technique for you.

**Actual titrations:** Refill the buret and record the initial volume of the NaOH solution. Open the stopcock and add the sodium hydroxide solution to the flask quickly until you are within 1-2 mL of the expected endpoint. Rinse the walls of the flask with a stream of deionized water from the plastic wash bottle, swirling well to mix the solution. Add sodium hydroxide to the flask dropwise until the pale pink endpoint is reached. Record the final reading of the buret and a qualitative description of the intensity of the pink color of the solution (very pale pink, pale pink, pink, bright pink). Calculate the volume of the NaOH solution used in the trial before continuing. Refill the buret and titrate the remaining two samples in the same manner. A trial that is "bright pink" should be repeated, as that color indicates that the endpoint of the titration was missed. *A minimum of three actual titrations* is necessary for the post laboratory assignment.

The volume of NaOH added should be consistent from trial to trial, as each trial contains the same quantity of unknown acid. *If the total variation (that is, the range) of the three volumes of NaOH added is greater than 0.15 mL, perform additional titrations until consistent results are obtained.* Consult your instructor if you are unable to stay within the acceptable limit after four titrations.

When you are *sure* you are finished with the titrations, drain any remaining NaOH solution from the buret into your waste beaker. Then, dispose of all solutions and the contents of your waste beaker into the "Acid-Base Waste" container provided in the laboratory. Rinse the buret and 5-mL pipet *thoroughly* with tap water, and then with deionized water. Return both to their proper places.

# Experiment 9

# Gas Laws

## PURPOSE

To explore the interdependence of pressure, volume, and temperature of a gas and to determine the relationships among these variables. To use the ideal gas law for determination of moles in a chemical reaction.

## INTRODUCTION

Matter can exist in three different physical states: gas, liquid, and solid. A gas fills the volume of the container that it occupies and exerts a pressure on the walls of the container. The most readily measurable properties of a gas are its pressure, volume, temperature and amount. The standard units of volume, pressure, temperature and amount are liters (L), atmospheres (atm), Kelvin (K) and moles (n), respectively. The mathematical relationships among these variables are called the *Gas Laws. Standard conditions* for a gas (referred to as STP) are defined as one atmosphere pressure and 273 K. Kelvin is an absolute temperature scale with K = 0 being the lowest temperature possible. Remember that: 1.00 atm = 760 mm Hg, K = (273 °C + °C) (1 K/1 °C), 1 $cm^3$ (cc) = 1 mL and that 1000 mL = 1 L.

### Part A: Pressure and Temperature of Air

In this part, you will explore the dependence of the temperature of the air in a bottle on the pressure. The volume of the gas and amount of gas are fixed. Before you start the measurements, predict the dependence of the temperature, in Kelvin, on the pressure, in atm, of air and write this in your laboratory notebook. Sketch a graph of temperature versus pressure consistent with your prediction.

## PROCEDURE

**Special Equipment:** 125-mL Erlenmeyer flask with a rubber stopper two Luer-lock fittings and one blue plastic valve, plastic tubing with two Luer-lock connectors, pressure sensor, temperature sensor, LabQuest2 with power cord, 1000-mL beaker, stirrer hot plate with stirring bar

**Materials:** ice/water

**Laboratory Safety:**

- Make sure that your safety goggles are in place during <u>all parts</u> of this Experiment

Figure 1 shows the experimental setup. Use the clear tubing to connect the first Luer-lock fitting on the rubber stopper to the gas pressure sensor. Attach a plastic valve to the second Luer-lock fitting and close the valve (the blue valve should be perpendicular to the valve stem when closed). Fill the 1000-mL beaker with approximately 150 mL of ice, and then add 400 - 500 mL of water and a stirring bar. Place the beaker containing the ice-water mixture on the stirring hotplate and turn the stirring on. Immerse the Erlenmeyer flask in the ice bath and clamp the neck of the flask to a ring stand. Be sure that the water level in the beaker rises to near the bottom of the neck of the flask. Add more water if necessary. Position the temperature probe with a clamp in the ice-water bath near the middle of the bottle so that the tip of the sensor is not touching the beaker. Finally, attach the free end of the plastic tube to the pressure sensor.

Figure 1

Make sure that the pressure sensor is connected to the LabQuest2 interface at the Channel 1 port (**CH1**) and the temperature sensor at the Channel 2 port (**CH2**). The power cord should be plugged into the **AC** adapter port while the other end is plugged into the power source. Firmly press in all the cable ends.

Switch on the LabQuest2 by pressing the power button on the top left corner of the device. Once the LabQuest2 has been turned on you should see a screen similar to Figure 2. The default settings have pressure in mm Hg and the temperature in Celsius. Change the units to measure pressure in atmospheres and temperature in Kelvin. To change units, tap the appropriate **Channel** (channel 1 or 2) and select **change units**. Select the appropriate units from the pull down menu as in Figure 3. You now need to change the settings. Tap **Mode** with the stylus and you should see a screen similar to Figure 4. Then tap the pull-down menu and select **Selected Events** and then tap **OK**. You should see a screen similar to Figure 5. You are now ready to measure the pressure of air in the flask at various temperatures.

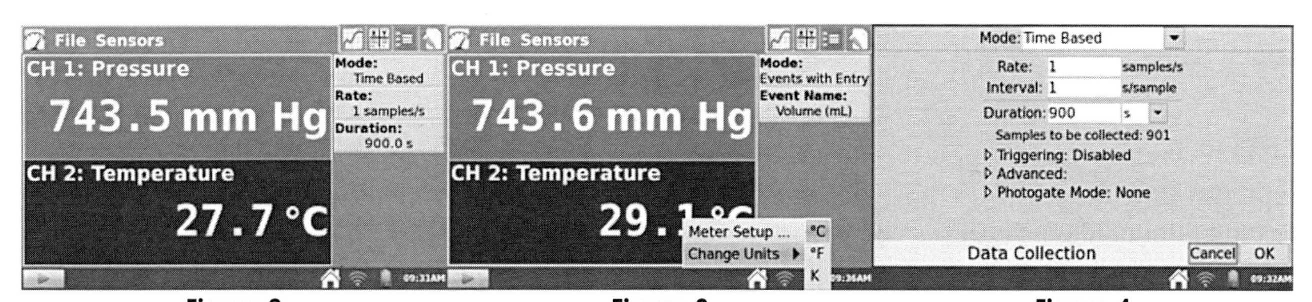


Figure 2 Figure 3 Figure 4

To collect data, tap the green start button (tap **Discard** if a message about overwriting previous data appears) and you should now see a screen similar to Figure 6. Allow the temperature in the ice-water to equilibrate with the temperature is in the range of 273 – 278 K. You can monitor the temperature on the right border of the screen.

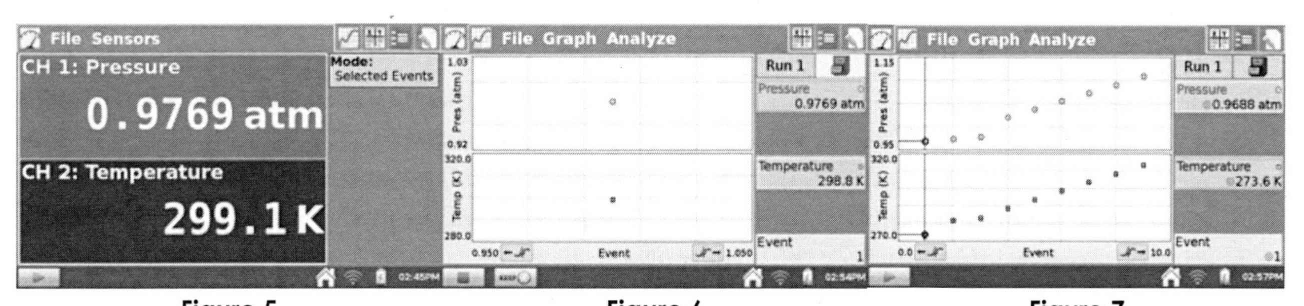


Figure 5 Figure 6 Figure 7

Before you start heating the ice-water bath, tap **KEEP** to store the initial pressure and temperature. Turn the heater control to 5 and heat the bath to 5 K higher than the initial temperature and tap **KEEP** to enter the second data set. Continue heating the bath, collecting data by tapping **KEEP** at approximately 5 K temperature increments as the water bath heats up to 315 K.

After you have entered the highest temperature, tap the stop button you should now see a screen similar to Figure 7 showing all the data points. Now tap the stop button (**Red Square**). Tap the **Table** icon to examine the data table. *Record all data in your notebook*. Tap the **Graph** icon to return to the graph screen.

### Part B: Application of gas laws to a chemical reaction

In this part, you will conduct a chemical reaction between solid magnesium and hydrochloric acid solution, as shown in the equation below.

$$Mg\ (s) + 2HCl\ (aq) \rightarrow MgCl_2\ (aq) + H_2\ (g)$$

You will react a known mass of magnesium with excess hydrochloric acid. The reaction will be conducted in a sealed vessel, thus trapping the $H_2$ gas formed. You will use the gas pressure sensor to measure the pressure increase in the sealed vessel, and a temperature probe to measure the temperature of the reaction indirectly, by measuring the temperature of the water bath in which the vessel is placed.

## PROCEDURE

**Special Equipment:** 100 mL graduated cylinder, 125-mL Erlenmeyer flask with a rubber stopper two Luer-lock fittings and one blue plastic valve, 20 mL plastic syringe, plastic tubing with two Luer-lock connectors, pressure sensor, temperature sensor, LabQuest2 with power cord, 1000-mL beaker, blue weighted ring

**Materials:** Magnesium ribbon, 1 M hydrochloric acid solution and water

**Laboratory Safety:**

- Make sure that your safety goggles are in place during all parts of this Experiment.

**Determining the Volume of a 125-mL Flask:** Obtain the 125 mL Erlenmeyer flask and rubber stopper that you will use for the experiment. The stopper should have the two Luer-lock fittings in place, with nothing attached to either fitting. Using tap water and a graduated cylinder, determine the volume of the flask available for the hydrogen gas to occupy during the reaction. Do this a minimum of three times, and calculate the average volume in liters available for the gas to occupy. Keep in mind the following:

- A 125-mL flask does not have a total volume of 125 mL.
- During the experiment, the flask will be sealed with the rubber stopper, which takes up some of the volume of the flask.
- You will added 5 mL of solution to the flask during the experiment.

**Set-up of Apparatus:** Figure 8 shows the experimental setup. Prepare a room temperature water bath in the 1000-mL beaker, with enough water to completely cover up to the neck of the 125-mL Erlenmeyer flask when the blue weighted ring is placed around the neck of the flask. Obtain an approximately 15 mm long piece of magnesium ribbon and record the mass to three decimal places. The mass should fall between (0.010 – 0.015 g). Place the magnesium in the clean, dry 125-mL Erlenmeyer flask. Seal the flask with the rubber stopper, ensuring that the stopper is securely in place so that the hydrogen gas will be trapped during the reaction. Place the weighted ring around the neck of the flask. Use the clear tubing to connect the first Luer-lock fitting on the rubber stopper to the gas pressure sensor. Attach a plastic valve to the second Luer-lock fitting and close the valve (the blue valve should be perpendicular to the valve stem when closed). Obtain ~ 20 mL of 1.0 M hydrochloric acid solution in a small beaker. Draw 5 mL of HCl solution into the 20 mL syringe. Attach the 20 mL syringe onto the two-way valve, ensuring that the numbers on the syringe are facing forward once the syringe is attached. Immerse the flask in the water bath (Figure 8). Be sure that the water level in the beaker rises to near the bottom of the neck of the bottle. Add more water if necessary. Position the temperature probe with a clamp in the water bath, ensuring that the tip of the sensor is not touching the beaker or the flask.

Figure 8

Connect the pressure sensor to the LabQuest2 interface at the Channel 1 port (CH1) and the temperature sensor at the Channel 2 port (**CH2**). Firmly press in all the cable ends.

Switch on the LabQuest2 by pressing the **power** button on the top left side of the device. Once the LabQuest2 has been turned on you should see a screen similar to Figure 2. Before you start collecting data, verify the units are set to measure pressure in atmospheres and temperature in Kelvin. If necessary, change the units as specified in Part A. Tap **Mode** on the opening screen and then tap the pull-down menu and select **Timed Based**, set the **Interval** to 5 s/sample and the **Length** to 360 as seen in Figure 9. Tap **OK** and you should see a screen similar to Figure 10. You are now ready to measure the pressure of hydrogen as the reaction occurs.

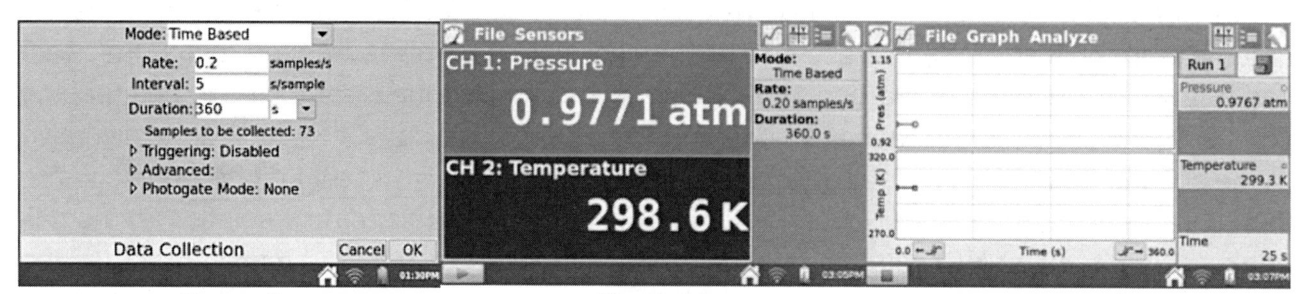


Figure 9 Figure 10 Figure 11

To start collecting data, tap the start button (**green triangle**) in the lower left corner of the screen. The temperature and pressure data will be collected every 5 seconds. The first few data points are used to determine the initial pressure and temperature. After approximately 20 seconds (elapsed time is found in the lower right hand corner of the screen (Figure 11), open the two-way valve directly below the syringe, press the plunger to dispense the 5 mL of HCl solution to the flask, and then pull the plunger back up to its original position. Close the two-way valve. As the reaction proceeds, bubbles should be visible indicating the generation of hydrogen gas. **Verify that the pressure is increasing with time**. *If the pressure is not increasing, stop the data collection because you have a leak in the apparatus set-up!* If the pressure is increasing with time, then the data collection will automatically stop after 360 seconds (Time window will return to 0 s). If the reaction is finished before that time (no more bubbles generated and no visible pieces of magnesium remain in the flask) you may click the stop button early.

Once the data collection ends, tap the **Table** icon to examine the data table. *Record the temperature, initial pressure and maximum pressure observed during the reaction in your notebook.* Remove the flask from the water bath, and carefully take out the stopper to release the pressure. Rinse, clean and dry the flask before starting each trial of the reaction. The solution that was removed should be discarded into the "Heavy Metals Waste" container. To start the next trial, tap the **File Cabinet icon** (Next to Run 1 in Figure 11). It will now say Run 2 and you can repeat the data collection process. A minimum of three trials should be performed.

# Experiment 10

# Evaluation of a Reactive Barrier for Pollutant Removal

## PURPOSE

To determine the calibration curve for a spectrophotometer and use this to investigate the effect of various treatments on the effectiveness of a model reactive pollutant barrier.

## INTRODUCTION

Contaminants affect the quality of surface waters such as streams, rivers and oceans as well as the quality of ground water. Ground water can supply drinking water through individual domestic wells and large community municipal wells. Additionally, ground water can enter streams and rivers, potentially transferring contaminants to surface water.

Organic compounds are a class of chemicals whose structure is comprised primarily of carbon atoms, along with lesser amounts of hydrogen, oxygen, nitrogen and other elements. Organic contaminants can enter groundwater through industrial activities, leaking underground storage tanks and improperly disposed waste materials. Once those contaminants are in the ground, removing them can be very challenging.

One technique that has been utilized for the removal of organic contaminants is called a *reactive barrier*. This technique is most easily applied in areas where shallow groundwater is contaminated. A trench is dug in an area where the contaminated groundwater is flowing underground and then the trench is filled with a material that will react with the contaminant, thereby removing it before it can flow further in the subsurface. This is shown schematically on the next page.

One material which can be used as a reactive barrier is elemental iron metal. In this experiment, you will set up an experiment to investigate the effects of elemental iron on the degradation of the synthetic dye indigo carmine. While indigo carmine is not itself toxic (it is used as a food dye, known in the food industry as FD&C Blue 2), it serves in this experiment as a model for an organic groundwater contaminant. To determine the concentration of this material, you will be performing what is known as a *spectrophotometric analysis*, in which the amount of light absorbed by a colored material is related to its concentration. The amount of light absorbed by a solution of unknown concentration will be compared with that of a set of

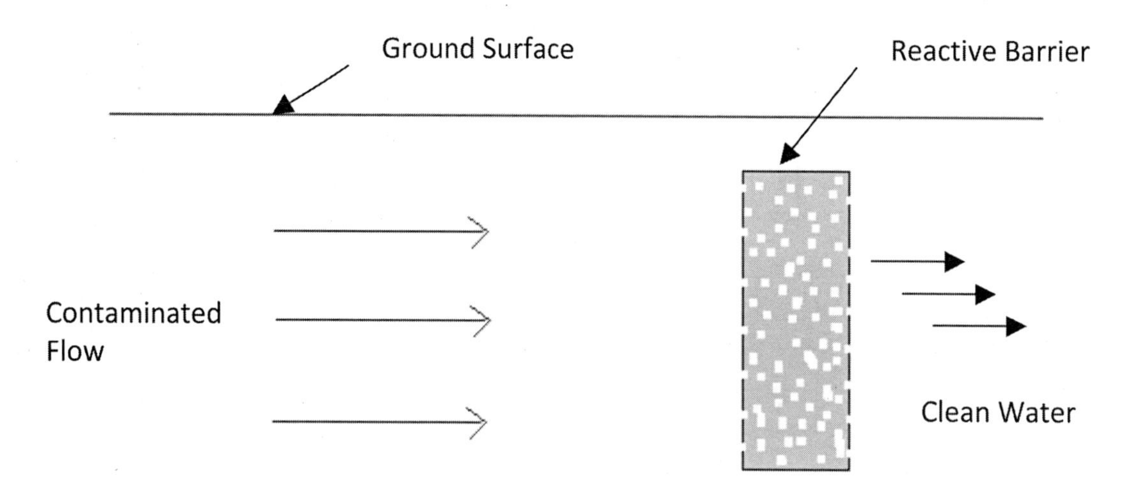


solutions of known concentration (that is, a set of *standard solutions*), and the concentration of the unknown can be determined. Refer to Appendix 12 for more information.

## PROCEDURE

**Special Equipment:** Centrifuge, shaker table, spectrophotometer, cuvets, 50-mL volumetric flasks, 25-mL volumetric pipet, 15-mL centrifuge tubes

**Materials:** Fe(*s*) powder, 20.0 mg/L indigo carmine (*aq*)

**Laboratory Safety:**

- Be sure that your safety goggles are in place during all parts of this experiment.
- Indigo carmine is not toxic but may stain skin and clothing.
- When using the centrifuge, be sure to *balance* the centrifuge rotor before turning it on (see below), and remember to allow it to come to a full stop before opening it and removing the tubes.

This experiment will determine whether indigo carmine contamination in groundwater could be treated using a reactive barrier made of elemental iron particles. You will be working with a partner for all parts of the experiment. Before beginning your laboratory work, discuss the procedure with your partner to determine the most efficient approach to the work. (That is, who will do what and when.) You should also plan the layout of the data tables in your laboratory notebook.

### Part A: Modeling Indigo Carmine Degradation in a Reactive Barrier

Obtain ten 50-mL plastic beakers, clean them if necessary, and shake out any excess water. Using a small piece of labeling tape, label each beaker with a code identifying your group and which sample number it is. Into beakers 3-6 (see the table below), weigh 1.0 g of iron (Fe) powder and record the masses. Try to keep the masses of Fe as consistent as possible from one beaker to another. Into beakers 7-10, weigh 2.0 g of Fe, again keeping the mass of Fe reasonably consistent from one sample to another and recording the masses used. Beakers 1 and 2 will not have any iron placed into them.

**Clean-up Balance Area: If you accidentally spill iron on the balance or the table, please sweep it up with one of the brushes available in the balance room. Ask your instructor for help if necessary.

Obtain about 90 mL of the 20.0 mg/L indigo carmine stock solution. Using the 15-mL centrifuge tubes, transfer about 8 mL of the indigo carmine solution into each of the ten beakers. Try to spread the Fe powder fairly evenly across the bottom of each of the beakers. Place your beakers on the shaker table operating at approximately 125 rpm. Beakers 1, 3, 4, 7 and 8 will stay on the shaker table for 10 minutes. Beakers 2, 5, 6, 9, and 10 will stay on the shaker table for 20 minutes. While the beakers are on the shaker table, you can proceed to Part B (but don't forget to take your samples off at the appropriate time!). *The mixtures need to be centrifuged and transferred to spectrophotometer cuvets (see the next paragraph) as quickly as possible after they are removed from the shaker table.*

| Sample Number | Reaction Time (min) | Mass Fe (g) |
|---|---|---|
| 1 | 10 | 0 |
| 2 | 20 | 0 |
| 3 | 10 | 1.0 |
| 4 | 10 | 1.0 |
| 5 | 20 | 1.0 |
| 6 | 20 | 1.0 |
| 7 | 10 | 2.0 |
| 8 | 10 | 2.0 |
| 9 | 20 | 2.0 |
| 10 | 20 | 2.0 |

After the appropriate amount of time has elapsed, remove the beakers from the shaker table. For samples 3-10, transfer the solution from the beaker into a 15-mL centrifuge tube. (The samples in beakers 1 and 2 do not need to be centrifuged and should be left in their beakers.) Be sure to label each centrifuge tube to correspond with the beaker number. After reading the instructions for use of a centrifuge in Part B of Appendix 6, centrifuge the tubes for approximately 2 minutes. Remember to balance the centrifuge, as shown below, and remember to allow it to come to a full stop before opening it and removing the tubes.

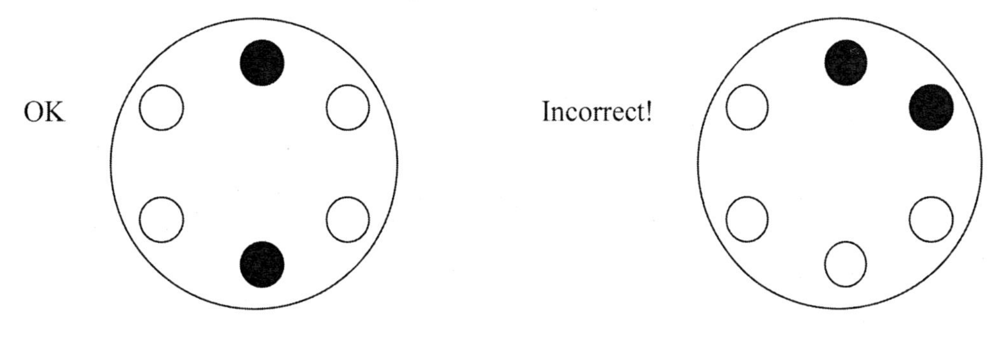

Read Section B1a of Appendix 12 on the use and care of cuvets. Obtain a set of 15 cuvets and rinse them with deionized water if necessary. (You only need ten of these cuvets at this point, but need the other five for the solutions prepared in Part B below.) Label these with appropriate labels using a small piece of tape near the *top* of the cuvet. Before filling them for analysis, they should be rinsed two or three times with a *small* amount of the solution being measured to flush out any water or other substances present (which would change the concentration of the solution). After each tube has been centrifuged, transfer the solution to a labeled cuvet using a dropper pipette. **Be careful to avoid transferring the Fe particles!** The cuvets should be filled approximately ¾ full (i.e., to within about 5 mm of the top). The solutions from beakers 1 and 2 can be transferred directly to the cuvets without being centrifuged. After preparing the standard solutions in Part B below, measure the absorbance of all of the solutions (blank, four standards, ten test solutions) using a spectrophotometer as outlined below in Part C.

Dispose of the excess indigo carmine solutions in the laboratory sinks and dispose of the Fe powder in the laboratory trash cans, **not** in the laboratory sinks.

### Part B: Preparation of Standard Indigo Carmine Solutions

While your samples are on the shaker table in Part A, you can begin to prepare the solutions required for Part B.

Obtain about 40 mL of the 20.0 mg/L indigo carmine stock solution in a clean beaker. In addition to what is needed for the preparation of the solutions, this should be sufficient for you also to have enough to rinse your volumetric pipet if necessary. You will use this solution to prepare several other indigo carmine solutions with lower concentrations using a process called *serial dilution.* Obtain three 50-mL volumetric flasks, rinse them thoroughly with deionized water, and shake out the excess water. Thoroughly rinse and dry their closures (stoppers or snap-caps, as appropriate). You will use these to prepare three standard indigo carmine solutions (that is, three solutions of known concentration) from the 20.0 mg/L standard solution provided.

1. Using a 25-mL volumetric pipet, carefully transfer 25 mL of the 20.0 mg/L solution to a 50-mL volumetric flask. Fill to the mark with deionized water, stopper or cap the flask, and invert several times to mix the solution, each time allowing the air bubble in the flask to rise completely to the top to mix the solution thoroughly and uniformly. This solution now has a concentration of 10.0 mg/L. Label the flask with the concentration.

2. Rinse the 25-mL pipet with a small amount of the 10.0 mg/L solution just prepared and discard the rinse solution. Now carefully transfer 25 mL of the 10.0 mg/L solution to a new 50-mL volumetric flask. Fill to the mark with deionized water and mix the solution thoroughly as described above. This solution now has a concentration of 5.0 mg/L. Label the flask with the concentration.

3. Rinse the 25-mL pipet with a small amount of the 5.0 mg/L solution just prepared and discard the rinse solution. Now use this to transfer 25 mL of the 5.0 mg/L solution to a new 50-mL volumetric flask. Fill to the mark with deionized water and mix the solution thoroughly as described above. This solution now has a concentration of 2.5 mg/L. Label the flask with the concentration.

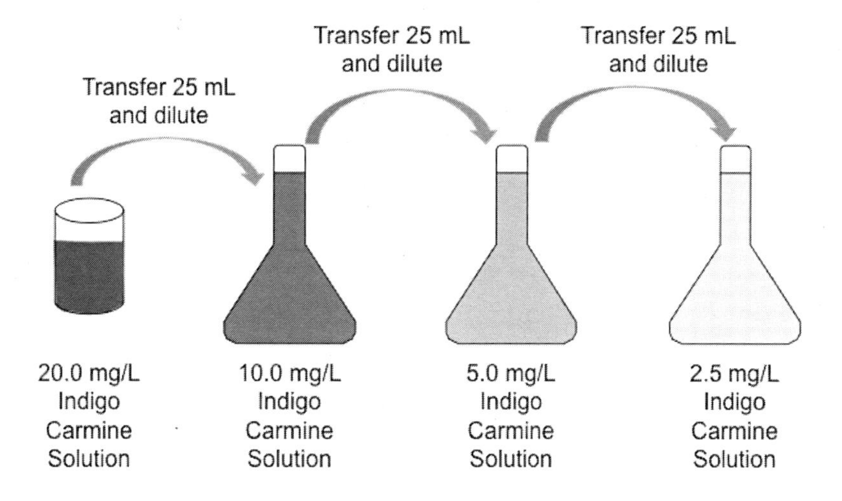


## Part C: Spectrophotometric Analysis of Indigo Carmine Dye

You probably already noticed that the intensity of the blue color decreases as the solution concentration decreases. In other words, the amount of color is directly related to concentration of the solution. Because we know the concentrations of the indigo carmine dye in these solutions, we can develop a mathematical relationship of the intensity of color as a function of the solution concentration. When this relationship is graphed, it is called a calibration curve or standard curve for the instrument being used.

Using a dropper pipet, transfer a sufficient amount of the 20.0 mg/L indigo carmine solution and each standard solution prepared in Part B above into your remaining cuvets, one solution per cuvet. As with the solutions from Part A, rinse the cuvet twice with a small amount of the solution to be measured before filling it about ¾ full with the solution for analysis. Don't forget to label the cuvets with the solution concentration.

At this point, you should have fourteen labeled cuvets containing indigo carmine solutions—four standards prepared in Part B and ten solutions from Part A. Fill your last cuvet with deionized water, which will be used as your blank solution (Appendix 12, Section B1b). Your instructor will specify which type of spectrophotometer you will be using. Read Section 2 of Appendix 12 which describe the use of a spectrophotometer to measure the absorbance of a solution. Measure the absorbance of your four standard solutions and ten solutions from Part B at a wavelength of 608 nm, and record the results in your laboratory notebook. When finished, the solutions can be poured into the laboratory sinks and flushed down the drain with water.

# Experiment 11

# Calorimetry: Heat of Reaction

## PURPOSE

To explore qualitatively the heats of solution of two solids, and using a calorimeter, determine quantitatively the heat of solution of a solid and the heat of neutralization of a strong acid with a strong base.

## INTRODUCTION

The definition of *chemistry* is often given as "the study of matter and the changes it undergoes." A release or absorption of energy almost always accompanies physical or chemical changes in matter. It follows that the forms of energy, their effects on matter, and the ways in which they can be produced from matter are all important concerns for a chemist. This experiment deals with some simple chemical changes and the energy involved.

Processes that release energy in the form of heat are said to be *exothermic* and those that absorb energy in the form of heat are said to be *endothermic*. The heat transferred in any chemical reaction is called its *heat of reaction*, which depends on the nature of the reactants and products. The amount of heat involved depends also on the amount of material reacted and is therefore an extensive quantity. When the *amount* of heat is expressed in units of joules (J) or kilojoules (kJ) per gram or per mole of reactant or product, it becomes an *intensive* quantity. In this experiment you will determine the ***molar*** *heat of reaction*; that is, the amount of heat involved per mole of reactant, for two processes.

The heats corresponding to specific types of chemical reactions have specific names, such as heat of combustion, heat of solution, and heat of neutralization. The *heat of solution* of a substance in a solvent is the amount of heat transferred (either released or absorbed) when a given amount of the substance is dissolved in a specified solvent. This energy will be expressed in joules or kilojoules. The ***molar*** *heat of solution* refers to the quantity of heat transfered *per mole of solute* and will be expressed in kilojoules per mole of solute (kJ/mol). The *heat of neutralization* is the amount of heat involved in the reaction between a certain amount of acid and a base. This energy will also be expressed in joules. The ***molar*** *heat of neutralization of an acid* refers to the quantity of heat involved *per mole of acid neutralized*.

In this experiment, you will explore qualitatively the effect of exothermic and endothermic reactions of a system on its the surroundings by examining the dissolution process of two solids. The terms *system* and *surroundings* are defined by the experimenter to fit a particular situation, but for a chemical reaction, the *system* most commonly refers to the reactant(s), and the *surroundings* refer to everything else surrounding the system (solvent, container, etc.). In the case of a reaction performed in a perfectly insulated calorimeter, the surroundings would be limited to the calorimeter itself (i.e., the container, stirrer, thermometer) and its contents (i.e., the reaction mixture and solvent) and would not extend beyond the calorimeter to the rest of the universe.

A common misconception is to think that when a substance undergoes an exothermic reaction its temperature must decrease, since heat is "lost" to the surroundings. It is important to note that the heat energy involved is not *thermal* energy of the reactant(s), but rather its stored (potential) *chemical* energy. In fact, in this case, a rise in temperature is observed because the temperature is sensed in the *surroundings* (by a temperature-measuring device or by your hand on the outside of the container), and the surroundings gain the heat lost by the *system*. In contrast, when a hot object cools down (also representing an exothermic process, but one that is a *physical* rather than a *chemical* change), indeed it is the *thermal* energy of the hot object that is transferred to the sur-roundings. Refer to your textbook for a more detailed discussion of these terms.

In this experiment you will also determine quantitatively the molar heat of neutralization of hydrochloric acid with sodium hydroxide (Equation 1):

$$HCl(aq) + NaOH(aq) \longrightarrow H_2O(l) + NaCl(aq) \quad (1)$$

If the pressure of the system is held constant (as in this experiment), the heat flow ($q$) for a physical or chemical change is equivalent to the enthalpy change ($\Delta H$).

In accordance with the First Law of Thermodynamics (also called the Law of Conservation of Energy), the quantity of heat released by a system must equal to the amount of heat absorbed by the surroundings, and *vice versa*. As mentioned previously, the energy transfer shows up as a temperature change in the *surroundings*. The mathematical *sign* of the heat flow ($q$) tells us the *direction* of the heat transferred. A positive sign indicates that heat is transferred *into* the system from the surroundings, and a negative sign indicates that heat is transferred *out* of the system to the surroundings. Thus, heat flow for the system ($q_{sys}$) and that for the surroundings ($q_{surr}$) must have equal magnitudes but opposite signs, as shown in Equation 2 below. (Note that the negative sign on the right side of the equation does not mean that $q_{sys}$ must be negative.)

$$q_{sys} = -\, q_{surr} \quad (2)$$

Most frequently, measurements of heat flow are made with a *calorimeter*. This device consists of an insulated container in which the process is allowed to proceed, a way of mixing the contents of the container, and a way of measuring the temperature change of the mixture (a thermometer or thermal probe). In our case, the calorimeter will consist of two polystryrene foam coffee cups, a lid, and a thermal probe. The initial temperature ($T_i$) of the mixture is measured, the process whose heat change is being measured is performed, and the final temperature ($T_f$) of the mixture is measured. The change in temperature ($\Delta T$), defined as shown in Equation 3

$$\Delta T = T_{final} - T_{initial} \tag{3}$$

reflects the change in temperature of the surroundings (the contents in the calorimeter and the calorimeter itself). The heat flow of the surroundings ($q_{surr}$) is equivalent to the sum of the heat flow for the solution in the calorimeter ($q_{soln}$) and the heat flow for the calorimeter ($q_{cal}$).

$$q_{surr} = q_{soln} + q_{cal} \tag{4}$$

In a chemical reaction, since the heat flow for the system ($q_{sys}$) is the same as the heat of reaction ($q_{rxn}$), we can determine the heat of reaction from the two terms q*soln* and q*cal* (Equation 5).

$$q_{rxn} = -(q_{soln} + q_{cal}) \tag{5}$$

The amount of heat energy needed to raise the temperature of one gram of any material by one degree is called the *specific heat capacity* of that material[1]. The relationship between the heat flow for a sample ($q$), the specific heat capacity of the sample ($C_s$), the mass of the sample ($m$), and the change in temperature for the sample ($\Delta T$) is given in Equation 6.

$$q = C_s \, m \, \Delta T \tag{6}$$

The heat flow ($q$) can be expressed in a variety of units. The SI unit of energy (or heat) is the *joule* (J). Thus, $q$ has units of J and $C_s$ has units of J/(g·deg), where "deg" can be in either units of Celsius or Kelvin.

In any calorimetric measurement, the calorimeter itself will release or absorb some heat. The relationship between the heat flow of the calorimeter ($q_{cal}$) and the temperature change ($\Delta T$) is

$$q_{cal} = C_{cal} \Delta T_{cal} \tag{7}$$

---

1. Note that this is sometimes referred to as "*specific heat,*" rather than "specific heat *capacity*." Potentially, this can be confusing, since "heat" and "heat capacity," while related terms, are not the same thing.

where $C_{cal}$ is the heat capacity of the calorimeter (often called the *calorimeter constant*) and $\Delta T_{cal}$ is the change in temperature of the calorimeter. Thus, the heat of reaction ($q_{rxn}$) in Equation 6 can be determined by applying Equations 6 and 7.

For exact measurements, the heat capacity of the calorimeter ($C_{cal}$) must be determined. This constant is the amount of heat required to raise the temperature of the calorimeter by one degree and has units of J/deg, and will be necessary in your subsequent calculations of the heat of reaction for Part B.

## PROCEDURE

**Special Equipment:** double polystryrene coffee cups with lid, digital or laboratory thermometer, hot plate, 100-mL graduated cylinder

**Materials:** anhydrous $CaCl_2(s)$; $CO(NH_2)_2(s)$; $HCl(aq)$ (approximately 1.0 M) and $NaOH(aq)$ (approximately 1.1 M)

**Laboratory Safety:**

- Be sure that your safety goggles are in place during all parts of this experiment.
- Dispose of any excess acid and base in the appropriate waste container.

Before *beginning the laboratory you should* set up one table in your notebook for each part of this experiment to organize the collection of your data.

### Part A: Qualitative Exploration of the Heat of Solution

As described in the Introduction, the *heat of solution* of a substance is the amount of heat either released or absorbed when a given amount of that substance is dissolved in water. This is not to be confused with the *temperature of a solution*, which refers to the temperature of a solution at a given moment. These can be distinguished by the difference in the units in which these two quantities are expressed. A heat flow would be measured in heat units (J, kJ, cal, etc.), while temperature is measured in °F, K, etc. In this part of the experiment you will qualitatively examine the heats of solution of two solids, to determine whether the process of dissolving each solid is exothermic or endothermic. You will focus on the change in temperature as each solid dissolves in water.

Add approximately 1 mL of deionized water to a small test tube. While nestling the bottom of the test tube in the palm of your hand (where temperature changes can be more easily detected), use a clean and *dry* scoopula (*large* spatula from your drawer) to add a large scoop (about 0.5 to 1 g) of solid calcium chloride ($CaCl_2$) to the test tube. (Note: Calcium chloride is highly *hygroscopic*, meaning that it absorbs water readily from the atmosphere. Tightly close the lid to the bottle of calcium chloride ***immediately***!) With a clean glass rod or small spatula, stir the mixture rapidly to aid the dissolution process. Record your observations. Are the contents of the test tube getting hotter or colder? Does this mean that the dissolution process of calcium chloride is exothermic or endothermic? Answer all these questions in your lab notebook and in your report. Dispose of the mixture in the laboratory sink.

Using the same procedure, examine what happens when solid urea ($CO(NH_2)_2$) is added to a test tube of water. (Be sure to use a clean and dry scoopula.) Include the answers to the same questions in your lab notebook. Dispose of the mixture in the laboratory sink.

## Part B: Determination of the Molar Heat of Neutralization

Neutralization is an important type of reaction that you have already encountered in Experiment 5 (Chemical Reactions) and Experiment 7 (Acid-Base Titration), and will encounter in later experiments as well. It is the reaction of an acid with a base. In this experiment you will measure the amount of heat transferred when aqueous solutions of hydrochloric acid and sodium hydroxide are mixed, and calculate the molar heat of neutralization (see equation 1).

Obtain a calorimeter setup (double polystyrene foam coffee cups with lid and thermometer) and assemble the apparatus as shown by your instructor. Check to see there are no holes in the cups. Record the exact molarities of the NaOH and HCl solutions from the labels on the bottles. Using a clean and dry graduated cylinder, measure 50-55 mL of the NaOH solution into the calorimeter cup and record the volume to the nearest 0.1 mL. Replace the lid and thermal probe on the cup and allow the assembly to stand for a few minutes to reach constant temperature. Using a clean and dry graduated cylinder, measure out 45-50 mL of the HCl and record its volume to the nearest 0.1 mL.

Record the temperature of the NaOH solution in the calorimeter with the thermometer. Using a separate thermometer, record the temperature of the acid solution in the graduated cylinder, to the nearest 0.1°C.

Quickly, but carefully, transfer all the acid from the graduated cylinder into the NaOH solution in the calorimeter. **Immediately** replace the lid and thermal probe on the calorimeter and swirl the cups gently to mix the contents, being careful to hold only the rim of the calorimeter.

Carefully observe the temperature of the calorimeter every few seconds. If the temperature is *rising*, note and record the temperature when it has reached a *maximum*. If the temperature is *dropping*, note and record the temperature when it has reached a *minimum*. It may take a few minutes for the temperature to show its greatest change.

Pour the solution from the calorimeter cup into the *acid-base waste container*, rinse with deionized water, and gently dry the cup. Perform a minimum of three trials. Rinse the temperature probe thoroughly with deionized water before returning it to its proper location.

## Calculations

Complete the calculations described below for each trial that you performed. The heat of neutralization, qneut, can be calculated using Equations 8 and 9.

$$q_{neut} = -(q_{soln} + q_{cal}) \qquad (8)$$

where $q_{neut}$ = heat flow for the neutralization
$q_{soln}$ = heat flow for solution
$q_{cal}$ = heat flow for calorimeter

and

$$q_{neut} = -(C_{s,\,soln}\ m_{soln}\ \Delta T_{soln} + C_{cal}\ \Delta T_{cal}) \qquad (9)$$

where $C_{s,\,soln}$ = specific heat capacity of the solution formed
$m_{soln}$ = mass of solution in calorimeter
$\Delta T_{soln}$ = temperature change for solution
$C_{cal}$ = heat capacity of the calorimeter = 20.5 J/°C
$\Delta T_{cal}$ = temperature change for calorimeter

We will assume here that the specific heat capacity of the solution formed is the same as the specific heat capacity of water. The mass of solution, $m_{soln}$, is calculated from the *total* volume of the solution in the calorimeter *after mixing* and the density (see Appendix 11) of the solutions at the initial temperature ($T_{initial}$). Assume that the densities of the HCl, NaOH, and product solutions are the same as for pure water. The initial temperature is the average of the measured temperatures of the HCl and NaOH solutions before mixing.

$C_{cal}$ is the heat capacity of the calorimeter apparatus. Although the textbook does not include this value in calculations of $q_{neut}$, in the real world, the coffee-cup calorimeter used does have a small, but measurable heat capacity. For this experiment, use a value of $C_{cal}$ = 20.5 J/°C. Finally, the change in temperature of the solution is $\Delta T_{soln} = T_{final} - T_{initial}$, and $\Delta T_{cal}$ is taken to be the same as $\Delta T_{soln}$. With these assumptions, calculate $q_{neut}$.

To determine the molar heat of neutralization (i.e., the heat flow per mole of acid neutralized), you must first calculate the number of moles of HCl and of NaOH that were mixed in the calorimeter. This is done by using the molarity of the HCl (recorded from the label of the bottle), the volume of the HCl solution used in the neutralization, and the corresponding molarity and volume of the NaOH. Remember that the molarity (M) is defined as the number of moles of solute per *liter* of solution. In the reaction, either the NaOH or the HCl could be the *limiting reactant.* Determine which reactant was in excess and which was the limiting reactant. It is the number of moles of the limiting reactant which determines the "number of moles of acid neutralized" to be used to calculate the molar heat of neutralization. From the heat of neutralization

($q_{neut}$) and the number of moles of the limiting reactant, the molar heat of neutralization of HCl (*molar* $q_{neut}$) can be calculated.

In the Calculations section of your laboratory notebook, using data, show setups for calculating $m_{soln}$, $\Delta T_{soln}$, $q_{soln}$, $q_{cal}$, heat of neutralization ($q_{neut}$), the number of moles of limiting reactant used in the neutralization, and finally, the molar heat of neutralization of HCl (molar $q_{neut}$). Since the experiment is performed under constant pressure, $q_{neut}$ and molar $q_{neut}$ are the same as the enthalpy change ($\Delta H_{neut}$) and molar enthalpy change (molar $\Delta H_{neut}$) for the reaction. Pay close attention to the signs and significant figures in your calculations.

Experiment 12

# Bonding and Molecular Structure

## PURPOSE

To draw Lewis structures and build models for molecules and polyatomic ions; to predict resonance and geometry.

## INTRODUCTION

In covalently bonded substances, the smallest unit that displays the properties of the substance is a *molecule*. These substances are therefore called molecular substances. Molecules are groups of covalently bonded atoms with an overall neutral charge. In ionic substances, the fundamental units are cations and anions attracted by ionic bonds. Many simple cations and anions consist of one atom which has lost or gained one or more electrons (to form a positive or negative charge, respectively), called a *monatomic ion*. However, a covalent species exists in many ionic substances—the *polyatomic ion*, a group of covalently bonded atoms with an overall negative or positive charge. In this experiment, you will examine both molecules and polyatomic ions. Given below is only a summary of the background and theory for the experiment; before coming to lab to perform this experiment, you should study the relevant sections in your textbook for more details. It would be a good idea to bring your textbook with you to the lab.

A convenient device for representing the bonding in covalent species is the Lewis electron dot structure, or simply, *Lewis structure*. A general procedure has been developed to translate the molecular formula of a species into the Lewis structure. Review that procedure from your textbook or lecture notes before coming to lab. In some cases, more than one Lewis structure can be drawn for a single formula—those species with isomers and those with *resonance structures*. *Isomers* are structures with the same molecular formula but different spatial arrangements of the atoms. They are structurally distinct species with different physical and chemical properties. *Resonance structures* are those having the same spatial arrangements of the atoms, but different arrangement of the valence electrons. The combination or average of the resonance structures (the resonance hybrid) is equivalent to the true structure of the species.

Valence Shell Electron Pair Repulsion Theory: The Lewis structure displays the number and arrangement of lone pairs and bonds in a molecule or polyatomic ion, but does so only in two dimensions. It does not necessarily represent the actual (i.e., 3-dimensional) molecular shape or geometry. However, predicting the correct Lewis structure is the key to determining *molecular geometry*. A simple approach is the Valence Shell Electron Pair Repulsion (VSEPR) theory, which leads us to the *VSEPR structure*. Unlike the Lewis structure, the structure predicted from VSEPR is a *three*-dimensional geometry, and therefore is a useful tool for predicting the properties of the species. According to this theory, the molecular geometry can be predicted solely by the repulsion of *electron groups* (sometimes also referred to as *pairs*) attached to each central atom either in a lone pair or in a bond.

The *number* of bonds between the central atom and the peripheral atoms is not considered at all. (That is, a double or triple bond is counted as one electron group.) The geometry of a species is determined by the number of electron groups around the central atom and whether these correspond to bonded pairs or to lone pairs.

**Note:** in the way the term is used here, a "central" atom is an atom which has at least two other atoms covalently bonded to it, regardless of its actual physical position in the structure.

## PROCEDURE

**Special Equipment:** Molecular models

**Materials:** None.

**Laboratory Safety:**
Not Applicable

You will be assigned species and told which specific items (Lewis structure, geometry, etc.) you are responsible for reporting. For each species assigned to you, start with the given **molecular formula**. (Note that this term is used whether the species is truly molecular (that is, neutral), or is a polyatomic ion.) Write the systematic name of the species, and work out a reasonable **Lewis structure**. Where appropriate, you will determine the Lewis structures of resonance structures as well. Based on the Lewis structure and structure type, you will apply the VSEPR theory and work out the **VSEPR** structure and/or structure type. You will then construct a model using a molecular model kit. The model will provide a physical representation that helps you visualize and remember the three-dimensional structure of the species. In particular it will help you name the **molecular geometry** (using the standard names provided in your text).

## Part A: Lewis Structures

Draw the Lewis structure(s) for each species assigned by your instructor. As discussed in the Introduction, your Lewis structure does **not** need to represent the actual geometry of the species, but does need to show correctly the number of lone pairs and clearly identify all covalent bonds.

## Part B: VSEPR Structures

Based on the Lewis structure, determine the structure type of the species and draw a VSEPR structure. Use your structure to make predictions about the electron group distribution, molecular geometry, and approximate bond angles. If there is more than one "central" atom, describe the molecular geometry at each of them. You may also wish to give an overall description of the molecule such as planar or linear if appropriate.

# Appendices

# Appendix 1

# Laboratory Safety

Safety in the laboratory should be a *constant* concern to all those involved—instructor and student alike. This does *not* mean that you should constantly be afraid that something awful will happen to you or that a chemical laboratory is an inherently unsafe place to be. Without doubt, accidents from burns, toxic chemicals, or broken glass can and do happen in chemical laboratories and injuries and even death can result from these laboratory mishaps. However, accidents of these types also can *and do* occur regularly in the kitchens and bathrooms of people's homes each year. Does this mean that kitchens and bathrooms are dangerous places? Accidents are most often caused by careless people, not by the situations in which they find themselves or by the materials they use. The primary causes of accidents, whether in the lab or at home, are careless technique, sloppy work habits, and lack of regard for those around you. All of these behaviors are easily avoidable if you exercise some caution about the operations that you are doing. The most general safety rule that can be given is: THINK. Know the properties of the materials being used and handle them accordingly. Be sure that you fully understand any procedure in its entirety *before* beginning it.

Early in the semester, you will be provided with a copy of the Safety Agreement below and asked to read it and return a signed copy to your instructor. The safety rules listed in the Safety Agreement are not meant to be an all-inclusive list of do's and don'ts for laboratory work. Rather, they are a sample of the types of concerns or attitudes that can bring about a safe laboratory experience. In general, proper behavior in a chemical laboratory is essentially common sense when working in an area where potentially dangerous materials or equipment are located: read labels carefully, no running or horseplay, clean up spills, don't taste chemicals, don't perform unauthorized experiments, use a pipet bulb or pump (no mouth-pipetting), be careful with flames around flammable materials, dispose of materials properly, turn off gas valves before leaving the lab, etc. The most important lab safety rules are these three:

1. **EYE PROTECTION MUST BE WORN AT ALL TIMES** when in an area where chemicals are being stored or used. This includes all times when you are in an area where other people are working, even if you are not personally handling chemicals. The policy at Towson University

is that you must have goggles which meet ANSI standard Z87 in place at all times when in the General Chemistry laboratory. In general, contact lenses are not considered safe for laboratory work, even if covered by glasses or goggles. The vapors or fumes from various chemicals can penetrate the material of the lenses and cause them to cloud over or possibly cause eye damage. Contact lenses should be removed prior to coming into the lab if at all possible.

**2. NO EATING, DRINKING, OR SMOKING IS ALLOWED IN THE LABORATORY.** As of August 1, 2010, smoking is not allowed anywhere on campus. Eating or drinking must be done outside the laboratory in the hallway. The possible dangers from fires or from accidentally ingesting materials while eating or drinking are too great to permit these three activities inside the laboratory.

**3. NO ONE (INCLUDING VISITORS) WITH BARE FEET IS ALLOWED IN THE LAB AT ANY TIME.** The recommended footwear is leather shoes that cover the tops of the feet (for protection from spills).

The exact penalty for violation of any of these three rules is up to your instructor. The first offense will generally result only in a warning. Second and third offenses will generally result in harsher punishment—expulsion from lab for some specified period, reduction in grade, etc., depending on the policy of your particular instructor.

**In summary:** *Knowledge* and *forethought* are your best tools when you are trying to achieve the goal of safe and efficient laboratory work. Read *all* the information provided to you *before* coming to lab, actively *think* about what you're doing and how to do it safely, and *ask questions* of your instructor if you're not sure of how best to proceed. And: if you or someone around you does have an accident, be sure to take the proper remedial action in as calm and rational a manner as possible to avoid further complications.

Guidelines for disposal of laboratory wastes are given in Appendix 2.

# GENERAL CHEMISTRY LABORATORY SAFETY AGREEMENT

Students are required to practice disciplined and responsible conduct at all times when present in the laboratory. Be alert and proceed with caution at all times in the lab. This safety agreement lists the lab-safety rules that are to be executed by everyone involved in order to ensure the safety of work place for everyone (students, faculty and staff). Each student must read, sign and return the agreement to their instructor.

## GENERAL RULES

1. Use of any type of food/drink (beverages, chewing gum, tobacco, etc), and cosmetics (lip balm, gloss) in the laboratory is prohibited.
2. Observe good housekeeping practices. Work areas should be kept clean and tidy at all times. Bring only your lab manual, lab notebook and other necessary materials to the work area. Keep aisles clear.
3. All written and verbal instructions are to be followed carefully. (If you do not understand a direction or part of a procedure, ask the instructor present before proceeding).
4. Unsupervised presence of students in the lab is prohibited.
5. Chemicals and equipment are NOT to leave the laboratory unless authorized by the instructor.
6. Fume hood sashes are not to be opened beyond the 18" mark when in use. (Never put your head into the hood.)
7. Hands and pens/pencils are to be kept away from face, eyes, and mouth while using chemicals or equipment. Hands are to be washed with soap and water after performing all experiments, especially before going to the restroom or leaving the lab for any reason.
8. Proper disposal of all chemical waste is a must. Label on the waste container must be checked thoroughly before adding chemical waste to the container. Waste containers are not to be overfilled. (Notify the instructor if the container is full.)
9. Sinks are to be used only for disposal of water and other solutions as specified by the instructor.
10. Know the location of, and how to use, the following safety equipment:
    a. Safety shower
    b. Eye wash
11. Report any accident (spill, breakage, etc.) or injury (cut, burn, etc.) to the instructor immediately, no matter how trivial.
12. Report fires to the instructor immediately.

## PERSONAL PROTECTIVE EQUIPMENT

1. Approved safety goggles MUST be worn at ***all*** times in the laboratory as indicated by your instructor. NO EXCEPTIONS.
2. Contact lenses are recommended to be replaced with prescription glasses.

3. *All skin must be completely covered from the torso down to the toes. Long pants and closed-toe shoes are required.* Any student wearing tank tops, shorts, capri pants (including athletic-wear), skirts*** or open-toe shoes will be denied permission to work in the laboratory.
4. Long hair, hanging items (jewelry, hoodie strings etc), and loose or baggy clothes should be secured.
5. Gloves are available for use when needed and must be removed before leaving lab.

***Exceptions will be made for religious observance. Please discuss with your instructor during the first week of laboratory!

## POLICY FOR STUDENTS WITH MEDICAL CONDITIONS

If you have any allergies, sensitivities, or medical conditions (diabetes, epilepsy, etc.) which might be aggravated by chemicals or potentially compromise your safety in the laboratory, be sure to inform your instructor in writing, *early* in the semester so that any special arrangements which may be necessary can be made for your protection.

Pregnant students should consult their physicians for advice on whether or not to perform experiments in the laboratory. Students are encouraged to provide their physician with a list of the chemicals that they might be exposed to while in lab. They should also check the MSDS sheets (available in the Department) to be aware of the hazards of the chemicals.

If a student is advised against performing laboratory work for a medically documented reason, then faculty must make accommodations for the student. Any accommodations should comprise a workload that is approximately equivalent to the regularly scheduled laboratory work. These accommodations may include:

- performing "dry" experiments only, in a place free from exposure to ongoing experiments;
- performing the wet chemistry at a later date;
- receiving an incomplete grade in the course pending completion of experimental work.

## SAFETY GUIDELINES ACKNOWLEDGEMENT

I have read the laboratory safety guidelines outlined above for this course in Towson University's Department of Chemistry and agree to abide by all the guidelines therein.

Printed Name______________________________________________________

Signature__________________________________________ Date__________________

Course Number_________ Section__________ Semester___________ Year_______

# Appendix 2

# Waste Disposal Guidelines

Because some of the materials used in the course have hazardous (flammable, toxic, carcinogenic, corrosive, etc.) or otherwise unpleasant (bad odor, irritating to skin or nasal passages, leave stain on clothing, etc.) properties, it is imperative that they be handled properly, including the proper disposal of excess materials at the end of a lab period. This is to ensure the safety and health not only of the persons working in lab, but also of the laboratory itself (plumbing, etc.) and the air, water, and land outside the building.

Towson University has a Department of Environmental Health and Safety, whose role is to give the campus community guidance on safety issues and handling wastes. This Department is also responsible for removal of hazardous waste from the Towson University campus. Our waste disposal guidelines are a combination of State of Maryland and Baltimore County guidelines.

General Chemistry laboratories will have, at minimum, two types of waste container. One container will be for wastes containing *heavy metals* and corrosives (generally non-flammable), and the other container will be for *organic* wastes (often flammable). In addition to these, there may also be containers labeled specifically for wastes from certain experiments or procedures. Students are directed to place waste chemicals in appropriately labeled containers, in accordance with guidelines from the Department of Environmental Health and Safety. These containers will have a label that says "Hazardous Waste."

Follow the directions given in the lab manual and by your instructor as to how to dispose of the chemical wastes from any procedure. When doing so, *carefully* read the label on the container before you put any waste into it, and put the waste in the specified container. Waste containers should not be filled to overflowing. If the container is full, inform your instructor so another container can be brought in for use. If you are not sure how to properly dispose of a particular waste, be sure to ask your instructor for directions *before* you dispose of it.

**Under no circumstances should waste chemicals be put into the laboratory trashcans or sinks unless you are *specifically* instructed to do so.**

# Appendix 3

# Accuracy, Precision, & Experimental Errors

## INTRODUCTION

Scientific experimentation is of two general types. First, *qualitative* observations can be made. These observations are made without reference to any numbers, scales, or unit systems. You might observe that when two substances are mixed, the mixture becomes warm. You might describe the color or texture of a material, or the size of the object as "large" or "small." In general, such observations can be of use in reaching some conclusions, but they are often considered to be preliminary—a prelude to the numerical measurements which will be used to reach more definitive conclusions.

Frequently, however, *quantitative* measurements are made during an experiment. Numerical val-ues are obtained for the magnitudes of various quantities. What is the maximum temperature (°C) reached by the reaction mixture? What is the volume of this object in mL? What is the mass of this object in kg? What wavelength (in nm) of light does this molecule absorb? In order to use the results of measurements properly, it is necessary to understand fully the meaning and limitations of these numbers, to know how to combine these numbers in a calculation to get the most meaningful and useful results, and to know how to account for and express the experimental errors in your measurements.

## ACCURACY AND PRECISION

The terms *accuracy* and *precision* are used almost interchangeably by some people, but actually mean very different things. The *accuracy* of a measurement or experimental result shows how close that result is to some "known," "correct," "true," or "accepted" value for that quantity. The *precision* of a series of measurements of the same quantity shows how closely that series of measurements agrees internally—that is, how close one measurement is to another within the set. The precision of a set of measurements is largely a function of how well that set of measurements was performed by the experimenter; in other words, it is a measure of the experimenter's technique. (This includes the experimenter's choice of equipment with which to make the measurement.) The accuracy of the measurement is also a function of the experimenter's technique, to be sure, but it is also related to the quality of the measuring instrument and the procedure that the experimenter uses.

To illustrate these concepts, the Figure on the next page shows four targets. The *precision* of the attempt to hit the bulls eye is measured by how closely the shots are grouped. The *accuracy* of the attempt is measured by how closely the "average" shot has come to the center of the target. Good agreement of the results with one another means that your technique was reproducible. If, however, that good precision was not accompanied by good accuracy, it is likely that the method (or, in this case, your rifle) needs adjustment or correction.

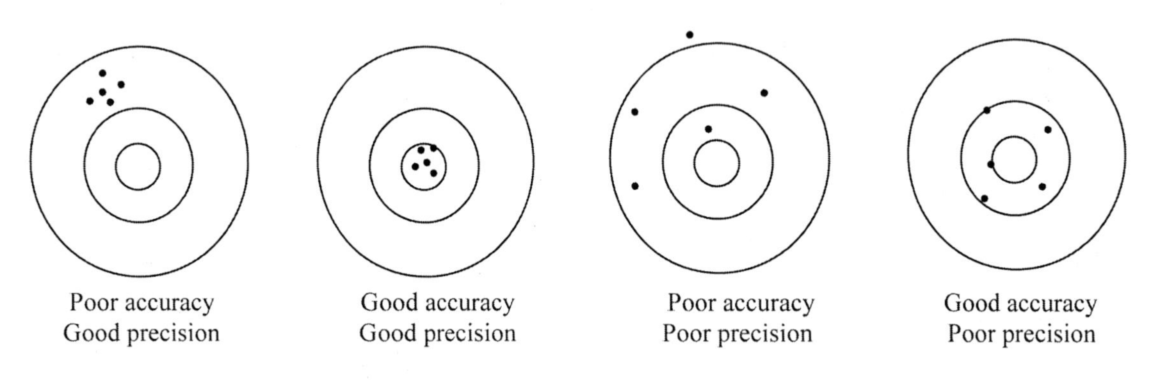


## EXPERIMENTAL ERROR

All measurements contain some amount of error or uncertainty. The process of measurement always involves the estimation of the last digit of the quantity being measured (or, it involves the inherent error or uncertainty of the measuring instrument if it gives a digital readout). Proper measurement of any quantity involves the reporting of all the digits of which you are certain, plus (exactly) one uncertain or estimated digit. These digits are referred to as the significant figures in the measurement. The rules for keeping proper track of significant figures are given in your textbook. These rules should always be followed when reading and recording experimental results, and in reporting results calculated from your data. When keeping track of the significant figures in a calculation, you should be aware that proper application of these rules does not just depend on the number of significant figures in the data. It also depends on the mathematical operation (addition, division, calculating a logarithm, etc.) being performed on that data. Be sure to study the appropriate section(s) of your text and apply these rules at all times. Honest reporting of the data and calculated results does not just include reporting the numbers; it also includes an (explicit or implicit) statement of the reliability and reproducibility of those numbers.

What is the source of these uncertainties or errors? The three general categories of error are instrumental, operator, and method errors. Instrumental errors are due to some inaccuracy in the measuring instrument: a poorly calibrated ruler or volumetric instrument, an electrical malfunction in a meter, a worn pivot in a balance. The user can sometimes "work around" the error (for example, a ruler which begins at 1 cm rather than at 0 can still give reliable results if the experimenter always remembers to subtract 1 cm from all the measurements), but usually the experimenter is limited by the measuring instrument.

*Operator errors* originate in the technique of the experimenter or operator of the instrument. Misreading the scale on a buret, spilling or contaminating a sample, failing to set the zero point on a spectrometer, and failure to clean a balance before using it are examples of operator errors. These are the types of errors over which the experimenter has the most control. If you think that you have made an operator error in a particular trial, this should be noted in your laboratory notebook. This notation may then enable you to reject that trial in reporting your results. (See the section "Rejection of a Trial" below.)

*Method errors* derive from a flaw in the method of analysis or measurement. In other words, the method used cannot produce accurate and/or precise results. A method that assumes that a material is 100% insoluble when it is really partly soluble will give poor results no matter what measuring instrument is used or who is using it. As with instrumental errors, the operator has limited control over this type of error, other than choosing an alternate method which gives better, more reliable results.

Errors can also be classified as *constant* or *random*. A constant error is always the same size and direction. A buret whose first mark after the zero is 2.00 mL instead of 1.00 mL will always give a volume reading that is 1.00 mL too large (assuming that you zero the buret each time), or a voltmeter may give a reading that is too small by a constant value due to some electrical circumstance in the system being studied. A random error may (conceivably) be of any size and of either direction. The fluctuation of a meter reading due to variations in the electrical current is a random error, as are the fluctuations in a balance caused by drafts deflecting the balance pan. Random errors can be treated by a statistical analysis of the results. (See the section below on statistics.)

## MEASURES OF ACCURACY

When comparing a result measured in the laboratory with a "known," "correct," "accepted," or "true" value, there is a variety of ways in which the error can be expressed. The *error* of the result (often referred to as the *absolute error*) is obtained by subtracting the known value from the measured value (*never* the reverse!), being sure to *retain the mathematical sign* of this difference. The *magnitude* of the error tells us how far away the measured value is from the known value, while the sign tells us in which direction the error lies. *A positive sign means that the measured value is greater than the known value, while a negative sign tells us that the measured value is smaller.*

By itself, the error in a measurement has meaning, but it can also be somewhat misleading. Is an error of –2 mg in a mass a significant error? If the object being weighed has a mass of 100 g, this error is insignificant, while if it has a mass of 5 mg, this error is quite significant. For this reason, the *relative error* is often computed. This is calculated by taking the ratio of the absolute error to the known value of the quantity being measured. Most often, this result is multiplied by 100 to give the *percent error.*

$$\text{Error} = \text{measured value} - \text{known value}$$

$$\text{Percent error} = \frac{\text{error}}{\text{known value}} \times 100 \qquad (1)$$

To keep proper track of significant figures, you may wish to calculate "error" separately and round it to the proper significant figures before calculating the percent error.

When the relative error is multiplied by 1000, we obtain the relative error in *parts per thousand* (ppt); when it is multiplied by $10^6$, we obtain the relative error in *parts per million* (ppm). The mathematical sign of the error is kept when the relative error is computed. By comparing the error with the value being measured, a more realistic picture of the magnitude (and potential importance) of the error is obtained. Note that the absolute error has the same units as the quantity being measured, while the relative error has no units (other than %, ppt, or ppm).

## PRECISION; STATISTICAL MEASURES OF PRECISION

Since there are so many ways in which errors can creep into any experimental measurement, it is highly unlikely for any set of repeated measurements to have exactly the same result in each trial. There can also be variability in the sample from one trial to the next. When a quantity has been measured repeatedly and several results obtained, we need to express the variability of the results in a way that can be easily understood by anyone interested in the overall reliability and quality of the measurements.

### Two Results Only

When only two results are being compared, the *percent difference* is often computed as a means of expressing the precision. This is calculated by taking the absolute value of the difference between the two results, dividing the result by one of the two values, then multiplying that result by 100. The mathematical sign on the difference and the percent difference is dropped, since you do not know which of the two values is the "correct" one (if, indeed, either of them is).

$$\text{Percent difference} = \frac{|\text{first result} - \text{second result}|}{\text{either result}} \times 100 \qquad (2)$$

If division by one result gives a substantially different percent difference than division by the other, you should then divide by their average, rather than by one of the two values. If this happens, however, it indicates that the agreement between the two trials is sufficiently poor that the experiment probably should be repeated to obtain a more reliable answer.

## More Than Two Results

With larger data sets, statistics provides several ways to express the precision of the results. In this discussion, consider a set of experimental results in which N trials were performed, all of which were supposed to give the same result. The individual results for trials 1, 2, ... will be denoted by x1, x2, .... The first calculation most commonly done is to find the *arithmetic mean*, or *average* (symbolized $\overline{x}$) of the results. This is calculated by taking the sum of the results and dividing by the number of trials N:

$$\overline{x} = \frac{\sum x_i}{N} \qquad (3)$$

(The Greek capital letter *sigma* (Σ) denotes the sum of whatever follows it.) The mean usually represents the most likely value of the quantity being measured. For example, the following set of data was obtained for a standardization of a “0.02 M” $KMnO_4$ solution: 0.01995 M, 0.01997 M, 0.02004 M, 0.02019 M, 0.02001 M. The arithmetic mean of these five trials is

$$\overline{x} = (0.01995\ M + 0.01997\ M + 0.02004\ M + 0.02019\ M + 0.02001\ M)/5 = 0.02003\ M$$

For very large data sets ($N \geq 20$), quantities other than the mean are sometimes used to express the most likely value for the quantity being measured. The *median* of a data set is that value which is in the middle of the range; that is, it is the value for which half the results are higher and half are lower. The *mode* of a data set is the result which occurs with a higher frequency than any other member of the set. Both of these quantities are not often used in typical sets of data generated in the laboratory, since acquisition of such large data sets is only very rarely done.

## Range

A simple measure of the precision of any set of results is the *range* (symbolized w) for the set of trials. The range is simply the difference between the highest and lowest values among the results. In our example, the range is 0.00024 M. This value does not really communicate any significant information about the precision. It is only when the range (or the other measures of precision being discussed here) are compared with the value of the quantity being measured that real meaning can be attributed to these quantities. The relative range of a series of measurements is equal to the range divided by the mean. In this case, the relative range is (0.00024 M)/(0.02003 M) = 0.012.

As discussed above for the error and difference, the relative value of any statistical measure of precision can be multiplied by 100 to give it as a percent, by 1000 to give it in ppt, or by $10^6$ to give it in ppm. In our example, the relative range could be expressed as 1.2%, 12 ppt, or 12,000 ppm.

## Deviation

A more meaningful measurement of precision is the *deviation* of each of the individual values. The deviation for trial i (symbolized $d_i$) is found by subtracting the mean of the set from the measured value for that particular trial:

$$\text{deviation for trial i} = \text{measured value for trial i} - \text{mean of values}$$

$$d_i = x_i - \bar{x} \qquad (4)$$

The mathematical sign is retained; note again that if $x_i < \bar{x}$ , the deviation for that trial is a negative number; values greater than the mean have positive deviations from the mean. If we average the *absolute values* of the deviations $|x_i - \bar{x}|$, we obtain the *average deviation* (symbolized d) for that set of results.

$$d = \text{average deviation} = \frac{|d_1| + |d_2| + |d_3| + \ldots + |d_n|}{N} \qquad (5)$$

$$= \frac{|x_1 - \bar{x}| + |x_2 - \bar{x}| + |x_3 - \bar{x}| + \ldots\ldots |x_N - \bar{x}|}{N} \qquad (5a)$$

where N again represents the number of trials. Both the individual deviations and the average deviation have the same units as the quantity being averaged. *Note: Do **not** average the values **with** their mathematical signs, since you would always get an average deviation of zero (or nearly so), and that does not necessarily mean you have good precision. Equations 5 and 5a are **not** the same as finding the absolute value of the average deviation.*

Dividing the average deviation by the mean gives the *relative average deviation* (rad) for that set of results. As with the relative range, the relative average deviation can be expressed in percent, parts per thousand, and so on.

Most frequently in this course, the relative average deviation in percent is used:

$$\text{Relative average deviation in percent} = \frac{\text{average deviation}}{\textit{mean} \text{ of the results}} \times 100 \qquad (6)$$

For the set of results given earlier, the deviations, average deviation, and relative average deviation are

$d_1$ = 0.00008 M  $d_2$ = 0.00006 M  $d_3$ = 0.00001 M  $d_4$ = 0.00016 M  $d_5$ = 0.00002 M

d = 0.000066 M

rad = 0.000066 M/0.02003 M = 0.0033 ≡ 0.33% ≡ 3.3 ppt ≡ 3300 ppm

Two significant figures have been kept in the calculation of the average deviation, even though only one would be justified based on the numbers. This practice of keeping at least two significant figures in the result of the calculation of the average deviation (and the standard deviation, discussed below) is generally accepted, even though, technically, it is not justified based on the rules for significant figures. In any calculation, any time that you drop to just one significant figure, a great deal of precision is lost. Speak with your instructor if you are not sure as to how to report these quantities, or how many significant figures would be appropriate.

Note that the range and the average deviation have the same units as the quantity being mea-sured, but the relative range and relative average deviation are unitless.

As a final measure of precision, the *standard deviation* is often calculated. This quantity has its greatest meaning when the number of replicates is large (N ≥ 20), but it can be used for smaller samples (down to N = 3) with some restrictions on its interpretation. The standard deviation (symbolized s) is calculated by taking the sum of the squares (**not** the square of the sum!) of the deviations for the individual trials, dividing the result by N – 1, and taking the square root of the result:

$$s = \sqrt{\frac{\sum (x_i - \overline{x})^2}{N - 1}} \qquad (7)$$

Technically, this formula gives the *estimated* standard deviation. The *true* standard deviation (symbolized by a lower-case Greek sigma, σ) can be calculated only for samples of infinite size. "Infinite size" is usually taken to mean N ≥ 20.

The standard deviation for any set of results is always larger than the average deviation. As with the range, the *relative* standard deviation can be calculated by dividing the standard deviation by the mean, and multiplying by 100, 1000, etc., if desired to obtain the result in percent, parts per thousand, etc. For our example, the standard deviation and the relative standard deviation are:

$$s = \sqrt{\frac{(-0.00008)^2 + (-0.00006)^2 + (0.00001)^2 + (0.00016)^2 + (-0.00002)^2}{5 - 1}}$$

$$= 0.000095 \text{ M}$$

$$\text{Relative standard deviation} = \frac{0.000095 \text{ M}}{0.02003 \text{ M}} = 0.0047 \equiv 0.47\% \equiv 4.7 \text{ ppt} \equiv 4700 \text{ ppm}$$

One of the most important uses of the standard deviation for a set of trials is that the statistical distribution of the results can be related to the magnitude of the standard deviation. For a large sample ($N \geq 20$), the distribution of results will be such that 68% of the results will be within one standard deviation of the mean, 95% will be within two standard deviations, and 99% will be within 2.5 standard deviations. For smaller samples, the statistics of this "normal distribution" (or "bell-shaped curve") will break down and these figures will apply only approximately.

## Rejection of a Trial

In a set of results, one or more results often seem suspicious in that they are quite far away from the others. (For example, you may have noticed that trial 4 in our data set is a bit farther from the average than the other trials.) Are there any circumstances that let us legitimately throw out these suspicious values so that they don't affect the average? If there is a *known* experimental problem with that trial, such as a solution being spilled, the wrong reagent added, an end-point overshot, etc., you would be justified in eliminating that result from consideration. This type of situation should be supported by an entry in your laboratory notebook. However, you would *not* be justified in acting as though that trial did not even exist. You should report *all* the data and calculations for that trial, and then exclude the results of that trial from the calculation of the average, average deviation, and so on, *and* include an explanation as to *why* they were excluded from the calculations. Totally ignoring the result and not reporting it at all is dishonest reporting of what you did in that experiment.

Suppose that no valid experimental reason exists for rejecting a result. Can you reject that trial just because it "looks wrong" or it "just can't be right"? Based simply on your "gut feelings," the answer is no! You must have a *valid* basis on which to justify rejecting a trial. It is possible, however, that random errors in the measurements (in the instrument, method, or operator) have compounded themselves in that trial to produce a result that is so far away from the mean that it is unlikely that the result is valid. In that case, you may be able to statistically reject a particular trial. There are statistical tests (4d test, t test, Q test, etc.) which can be done, but these calculations are beyond the scope of this course and will not be presented here. Consult a text on statistics or on Analytical Chemistry if you wish more information on these tests and their use. The general rule on rejecting a trial is that you should *discuss the situation with your instructor* for guidance.

## Considerations for Choosing Samples, Numbers of Samples, and Sample Sizes

As discussed above, any measurement is subject to errors of various types. For this reason, it is important that you plan your measurements with care, choosing your measuring tools and experimental methods to minimize error. Choosing your sample is every bit as important. How many trials should you run? How do you select your sample? What size should each of your samples be? The subject of *sampling* is an important one—but one which can be discussed only briefly here due to its complexity.

In general, multiple trials are run for laboratory measurements. How many replicate samples are "enough?" This is a question that does not have a simple answer. For a measurement such as mass, with a low % error, replicates are not that important. If the measuring instrument has a greater inherent relative error, then perhaps more trials are in order. It also depends on your skill level with the measurements you are making. More trials will be necessary if the technique is relatively new to you, but you can probably get along with fewer samples if you are an "old hand."

If the material being tested is homogeneous (which is usually the case in General Chemistry, but may not be the case in advanced-level courses or in "real life" after graduation), a small number of replicate samples probably suffices. If the material is non-homogeneous, you must choose your samples with care, and probably in larger number. Thorough mixing of the larger sample is necessary prior to removing your samples to ensure that the samples you choose for analysis are representative of the larger sample as a whole. Even after thorough mixing, you may wish to sample different parts of the sample (top, middle, bottom) to help ensure that you have obtained a random, representative sample to analyze.

If you are sampling an inherently non-homogeneous population, as is the case with most biological systems (are any two individual organisms ever *really* identical?), you must pay careful attention to the sampling protocol and will probably need to perform analyses on many more samples than in a typical chemistry experiment, where the "populations" are much more homogeneous and predictable. As a result, the "acceptable" level of variation in results is generally larger for biological samples than for chemical samples, due to the inherent non-homogeneity of biological populations.

And, lastly, what do you need to take into consideration about the sample size? Do you need to take a 1-gram sample, or will 0.1-g samples suffice? If you take *too small* a sample, your relative error may become unacceptably large. For example, if your balance reads to the nearest mg and your sample weight is 10 mg, then your relative error is 10%. Using the same balance, a 1-g (1000-mg) sample can be weighed with a relative error of 0.1%. Then is it always true that the larger the sample, the better? Not necessarily—there are times when a sample can present difficulty if it is *too large*. If your sample is a very expensive or rare one, then cost or availability may be a limiting factor, both in the number of samples run, and in sample size. The other complication from an excessively large sample can be in exceeding the capacity of the measuring instrument you are using. For example, if the sample of acid measured for an acid-base titration is too large, then the volume of base which needs to be added from the buret may exceed the maximum volume of the buret. Similarly, the amount of unknown in a spectrophotometric analysis may be such that the absorbance of the resulting solution is too large to be measured accurately.

# Appendix 4

# Using Excel® for Graphing (PC)

## A. INTRODUCTION – SOME BASICS ABOUT GRAPHING

A *graph* is essentially a pictorial way of presenting a set of data. This can be done in a variety of ways (such as a circle or "pie," or a bar graph), but the most common type of graph used in science is a *line graph*, where each point in a two-dimensional space is designated by an ordered pair of numbers. By plotting the points of the data set, it is often possible to discover or interpret the relationship between the two quantities being plotted. Conventionally, the axes are drawn perpendicular to one another. The horizontal axis is conventionally called the *abscissa* or *x*-axis, while the vertical axis is conventionally called the *ordinate* or *y*-axis. Conventionally, the *x* coordinate of the point is specified first. The point where the two axes cross (with coordinates (0,0)) is called the *origin* of the graph.

In choosing how to plot the data (i.e., which quantity is plotted on which axis), the convention is to plot the *independent variable* (i.e., the one over which the experimenter has control) on the *x*-axis and the *dependent variable* (the one which is measured in response to changes in the independent variable) on the *y*-axis. This means that changes in the independent variable are thought to be the *cause* of the changes in the dependent variable. Mathematically, we say that the dependent variable *is a function of* the independent variable. If you are asked to "plot A *vs.* B," or "make a graph of A *vs.* B," that means that quantity or variable A should be graphed on the *y*-axis and quantity or variable B on the *x*-axis.

Usually, the points on a graph are connected by some sort of *smooth* curve (which could be a straight line; to a mathematician, *any* line on a graph is called a curve), not a point-to-point, "connect-the-dots" sort of broken line. If this is the case, this indicates the existence of some sort of trend or regular relationship between the variables, as opposed to a group of unrelated random or isolated events. This would allow you to predict the results of an experiment at points between data points (*interpolate*), and to predict the results of an experiment at points beyond the range of the measured data (*extrapolate*).

In this course, *all* graphs submitted as part of a lab report or lab assignment must be generated using a computer graphing program. There are many different programs to choose from, but one of the most common is Microsoft Excel®, which is available on the public computers on campus. What is presented here is a very introductory guide to using Excel® (based on Excel® 2013) to produce graphs. This should give you the basic information and skills necessary to complete assignments in General Chemistry lecture and laboratory. This program is extremely powerful and sophisticated, with capabilities far beyond those required for this course or discussed below. If you haven't used Excel® before, consider this a good first step in learning how to use this program for this purpose.

## B. USING EXCEL® TO PRODUCE A GRAPH

### Entering Data

The most common way to organize data in Excel is by creating a vertical column with a label in the first row.

Example: Solutions of varying concentration are analyzed by an instrument which generates a signal based on the solution concentration.

| | A | B |
|---|---|---|
| 1 | Concentration (mol/L) | Signal |
| 2 | 1 | 0.009 |
| 3 | 2 | 0.021 |
| 4 | 5 | 0.046 |
| 5 | 10 | 0.095 |
| 6 | 20 | 0.213 |

In this case the independent quantity is Concentration and the dependent variable is Signal (we know the concentration, but we have to measure the signal). This means that Concentration should be on the x-axis and Signal should be on the y-axis. In Excel, the x-axis values need to be to the left of the y-axis values.

## Making a Line Graph ("Scatter Chart")

For starters, Excel refers to graphs as "charts". By graphing this data, we can see the relationship between signal and concentration.

1. Select the data to be graphed by clicking and dragging the cursor over the range of cells containing your data (including labels).
2. Select the **Insert Tab** from the Ribbon.
3. Select the **Scatter button** in the **Charts** group and select the first subtype (Scatter with only markers).

| | A | B |
|---|---|---|
| 1 | Concentration (mol/L) | Signal |
| 2 | 1 | 0.009 |
| 3 | 2 | 0.021 |
| 4 | 5 | 0.046 |
| 5 | 10 | 0.095 |
| 6 | 20 | 0.213 |

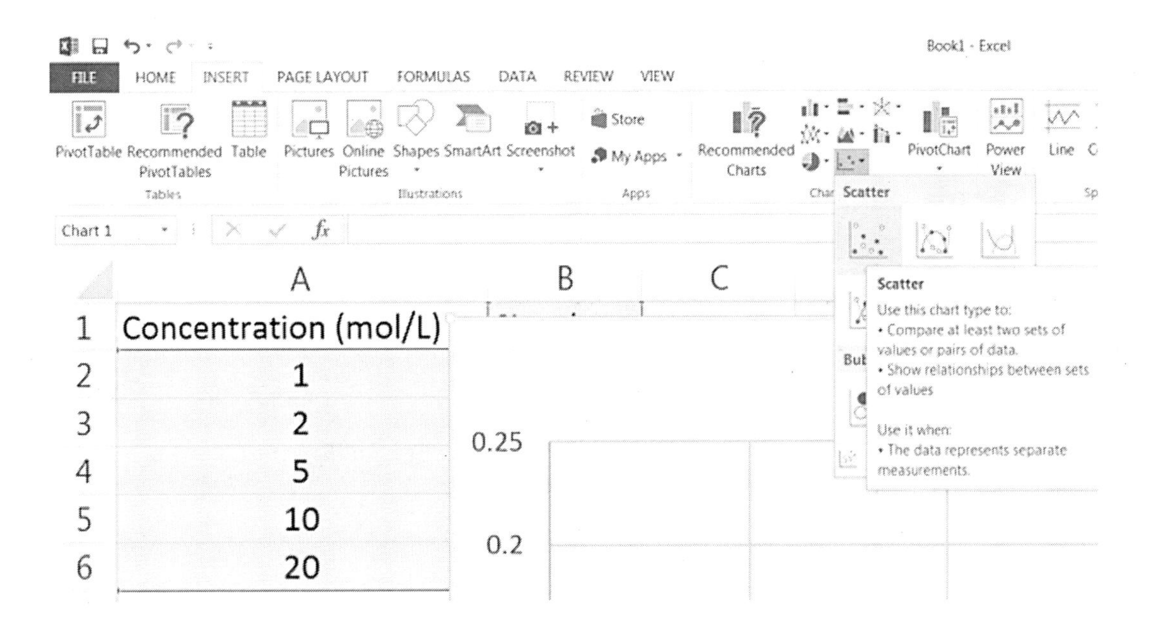

4. If the XY Scatter Chart is covering the data, move it by placing your mouse in a blank area of the chart until your mouse turns into a four headed arrow, hold down the left mouse button and drag the chart to a clear area on the worksheet.

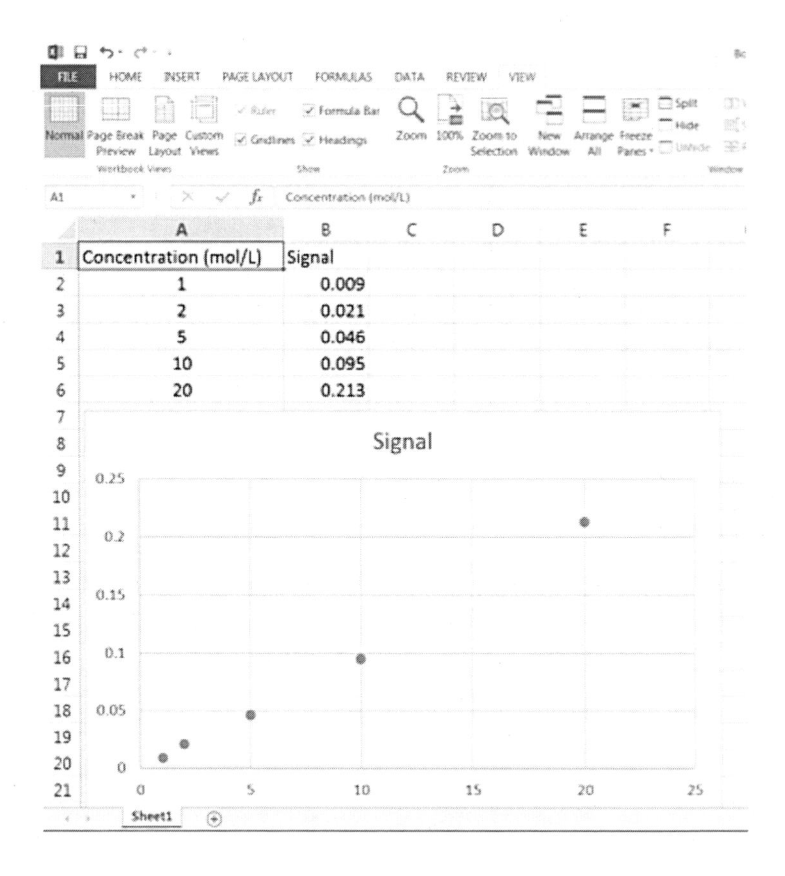


## Changing the Chart Title

1. Select the "**Chart Title**" above the chart itself and type directly over it with your desired title.

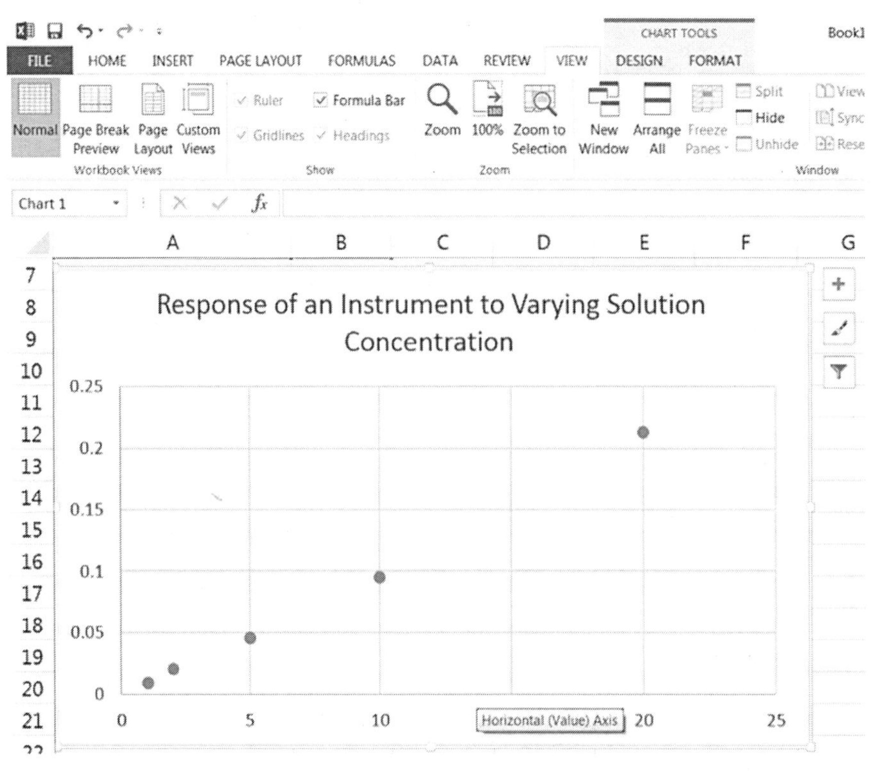

If there is no title:

2. Select the **plus symbol** to the right of the chart.
3. Check the **Chart Title** box.
4. Select the **Above Chart** option. The Chart Title text box appears above the chart.
5. Type your chart title and press ENTER on your keyboard.

**Note:** Depending on the length and complexity of the title you select, you may have to adjust the size of the overall chart borders, the size of the chart itself, and the relative positions of the chart and the title box. This is done by clicking one edge or corner of the box you wish to size or move and dragging it with the cursor. You can also click the "Format" tab, then use the "Size" group menu for this purpose.

## Adding Axis Titles

Axis labels should include the quantity being measured (data type) and the units (IMPORTANT!).

1. Select the **plus symbol** to the right of the chart.
2. Check the **Axis Titles** box. Both vertical and horizontal axis title boxes appear on the graph.

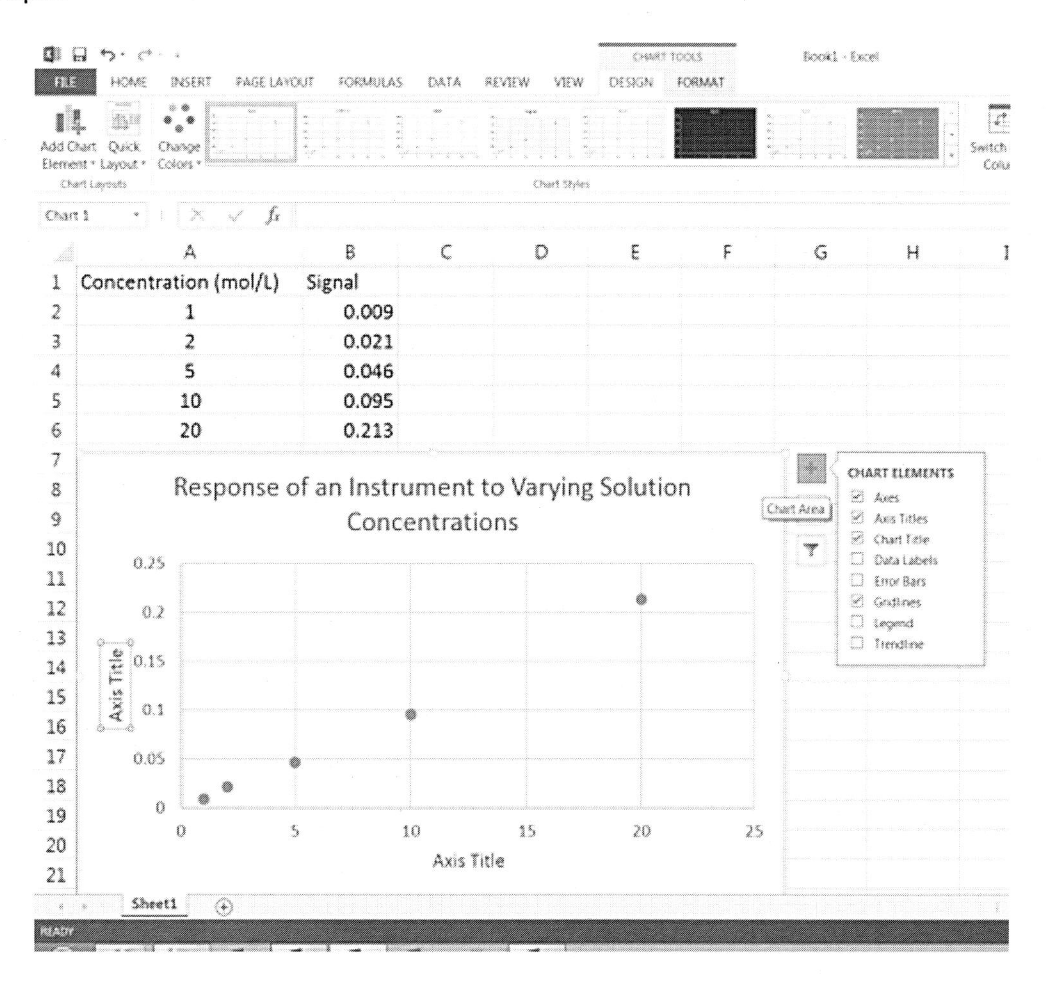

3. Type the vertical axis title you want (Signal (units) was used in this example) and press ENTER.
4. Type the horizontal axis title you want (Concentration (mol/L) was used in this example) and press ENTER.

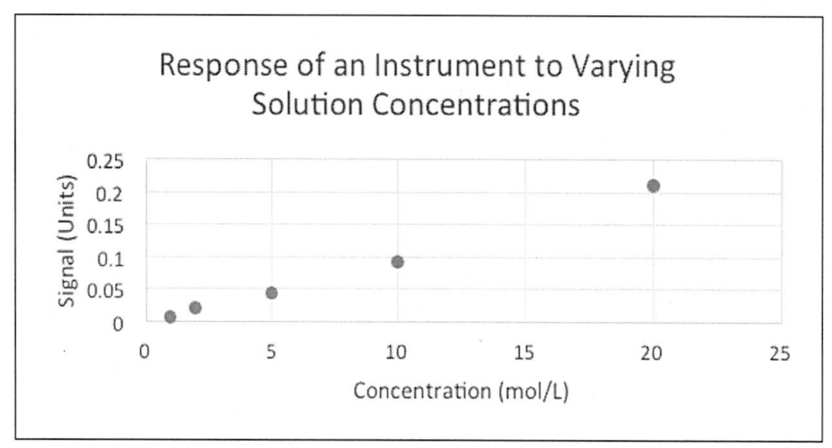


## Best Fit Line, Slope and Intercept

Often, we need to determine the mathematical relationship between the points on a graph. In the graph above, you can see that the points form a straight line. A best fit line can be added to the graph by the following the steps below:

1. Click once on the chart to select it.
2. Select the **Plus Symbol** to the right of the chart.
3. Check the **Trendline** box and click the triangle arrow next to the word Trendline.
4. Then select **More Options**… at the bottom of the list.

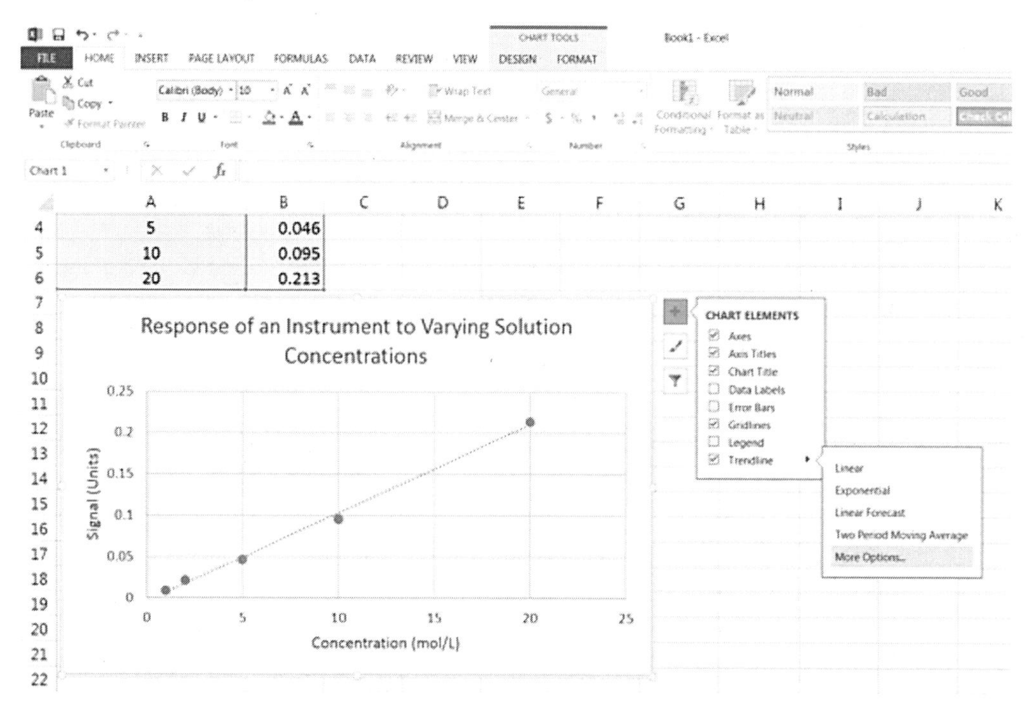

5. The **Format Trendline** dialog box will appear.
6. Click beside **Linear**. The equation of that line will follow the form: y = mx + b where m is the slope and b is the y-intercept.
7. At the bottom of this dialog box, check the box for **Display Equation on chart**. You may also want to check the box for **Display R-squared value on chart** for some graphs.
8. Click **Close (X) button** in upper right hand corner of the Format Trendline box to close.
9. When you return to your graph, it will have the equation of the best fit line, along with the R-squared value. Often the equation is in the middle of the graph and may be covering data points, so you can move it to one side by dragging the equation.

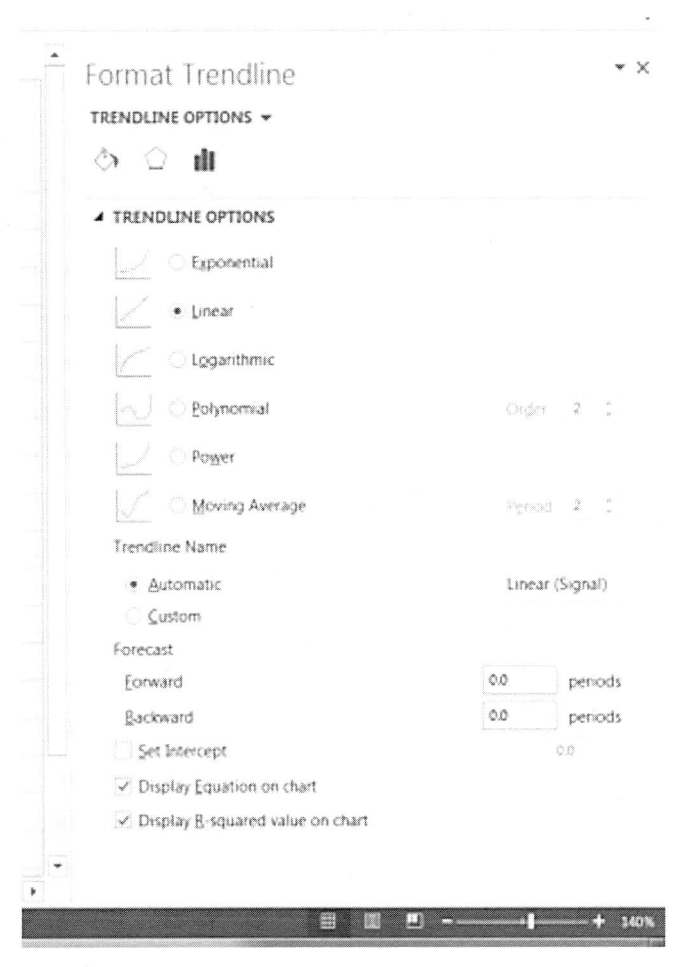


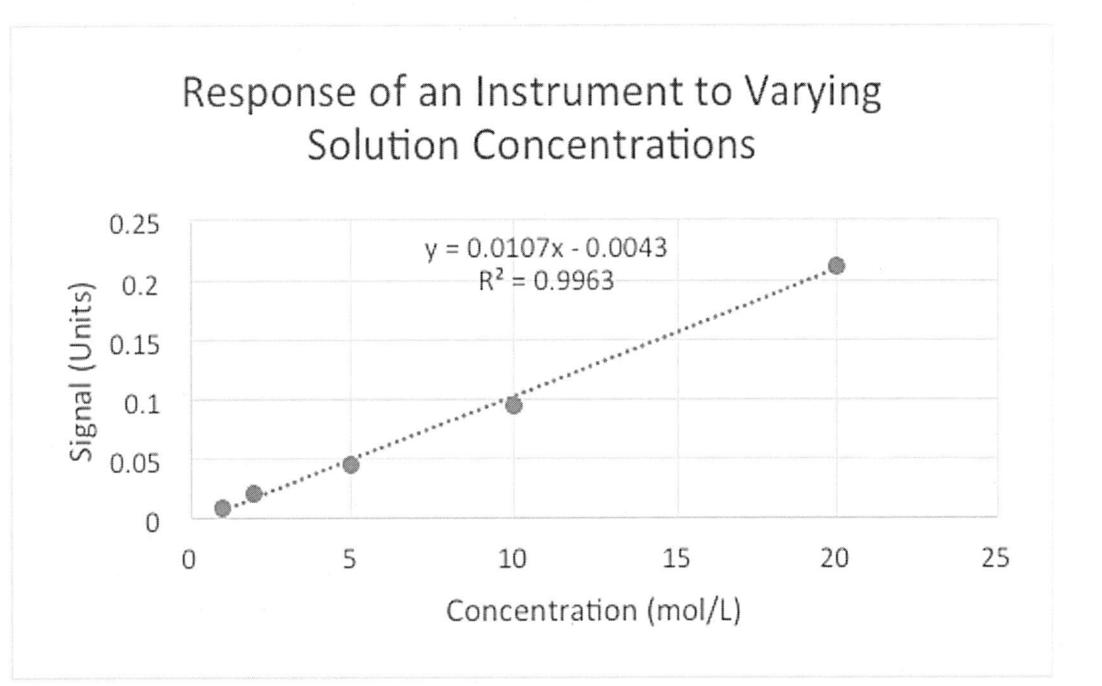

The R-squared statistic is a rough measure of how well the data fit the model of a straight line. The closer this value is to 1.0, the better the fit of the data to the linear model. In this graph, the points are all very close to the best fit line, so the R-squared value is high (above 0.99). An R-squared statistic below 0.9 is typically an indicator of a poor fit, and might prompt you to look for an outlier in the data points.

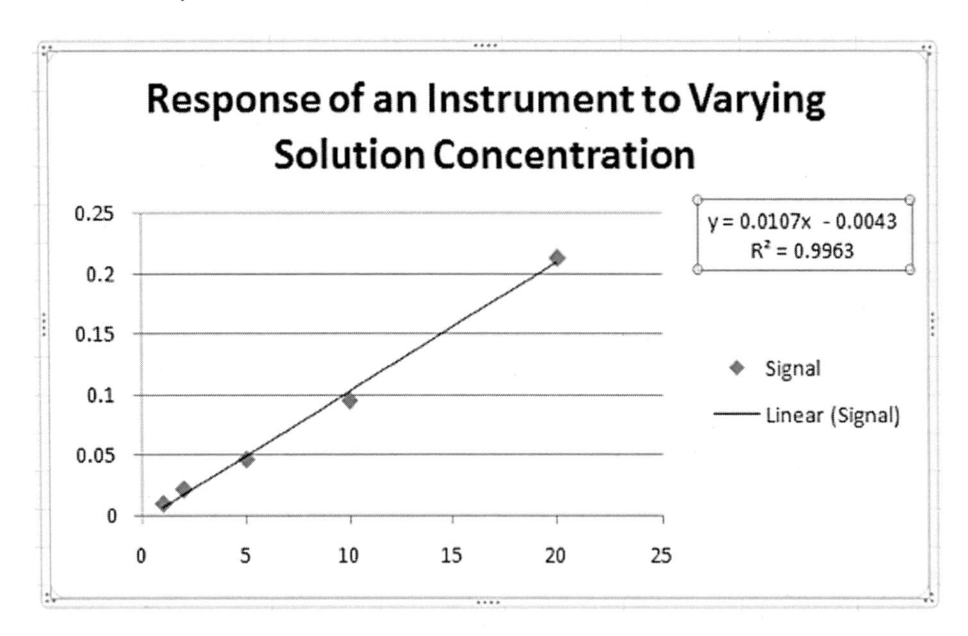


### Fine-Tuning and Printing

When printing, *you must make sure the graph is selected.* When selected, the graph will have a border around it and the 'Chart Tools' become available on the top ribbon. When the graph is not selected these 'Chart Tools' go away.

### Changing numbering on an axis

1. Click once on the chart to select it.
2. Select the **Plus Symbol** to the right of the chart.
3. Next to the **AXES** box, click the triangle arrow.
4. Then select **More Options**... at the bottom of the list.
5. The Format Axis box will appear on the right of the screen. The axis selected (x or y) is indicated by which axis has a box around it on the graph. Click on the other axis to move between them. In this example, the x-axis is selected, and so all of the values in the Format axis box relate to x-axis scale. Often the 'Auto' function does a good job of selecting the **Bounds** (range of the axis) and **Units** (increments where numbers are displayed for axis). You can type over any of the values and hit enter to change the way the data are displayed.

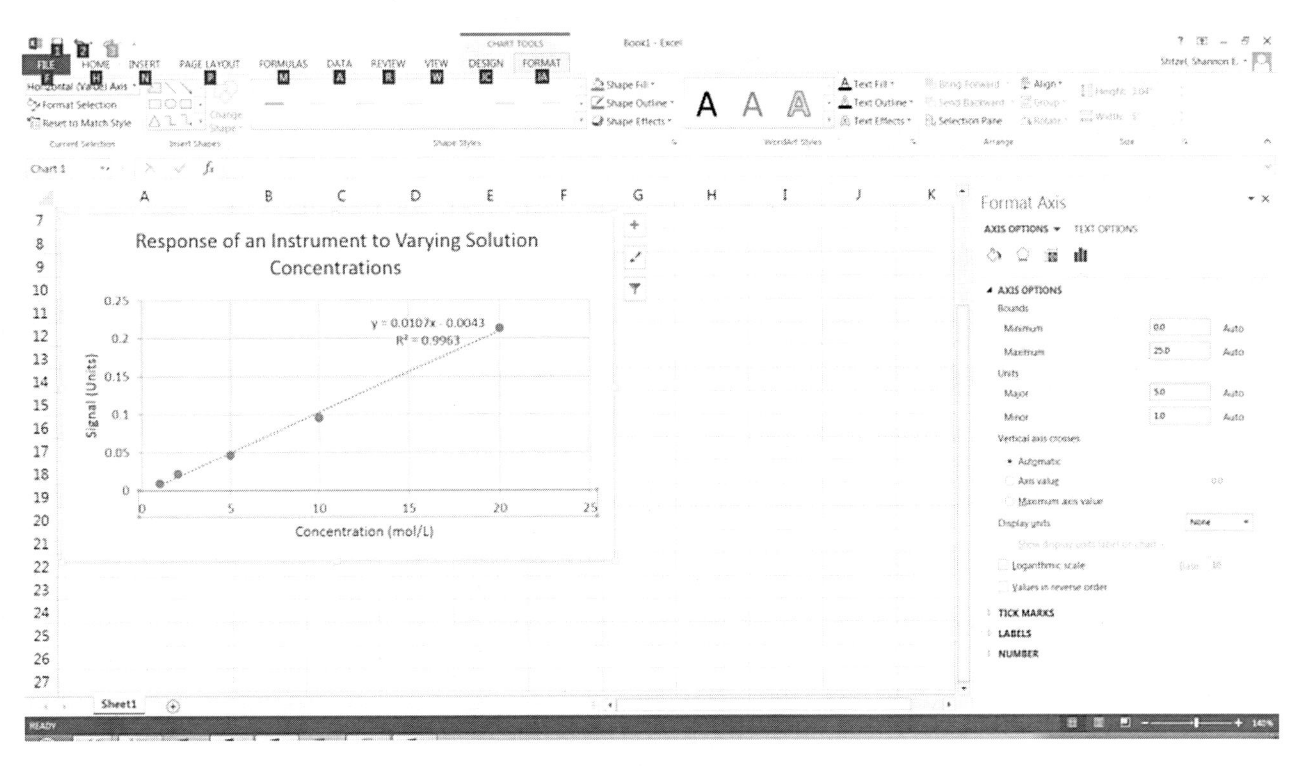


## Gridlines

1. Click once on the chart to select it.
2. Select the **Plus Symbol** to the right of the chart.
3. Next to the **Gridlines** box, click the triangle arrow.
4. Turn gridlines on and off by checking or unchecking the boxes as appropriate.

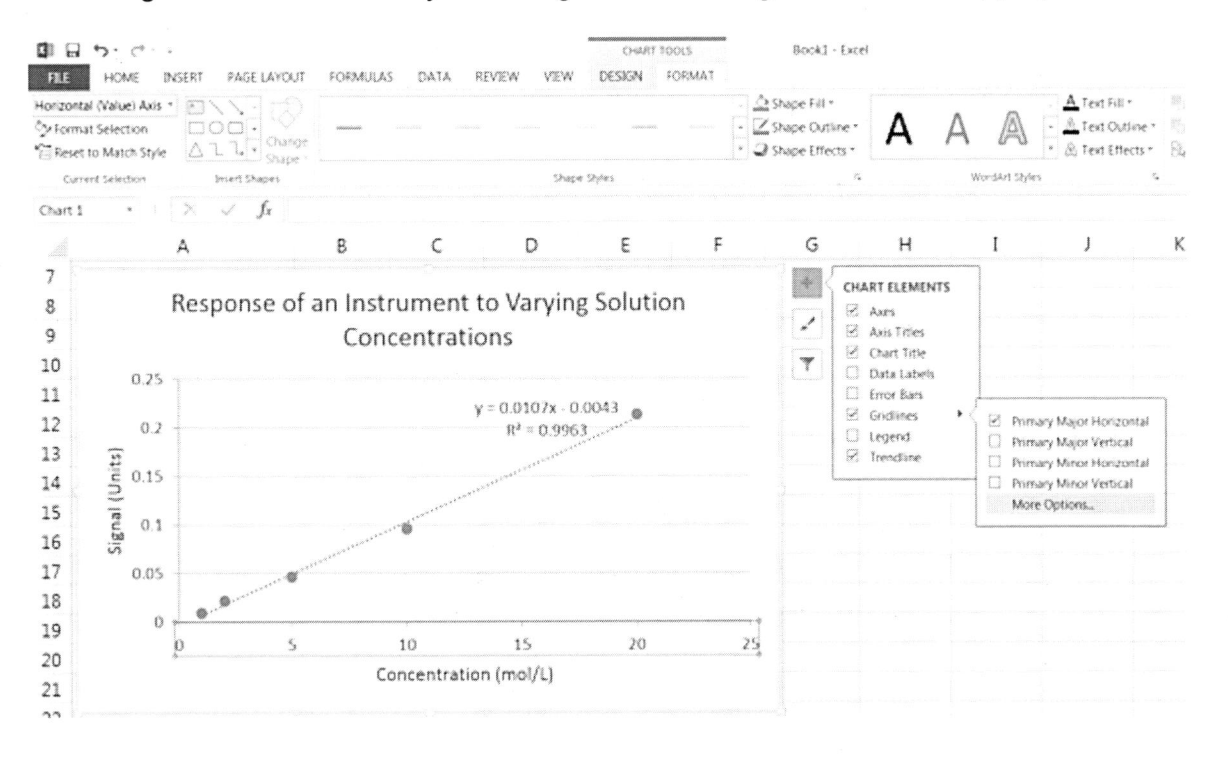

### Printing

1. To print the graph alone, on its own page, select it on the worksheet.
2. Click the **File** tab on the ribbon.
3. Select **Print**.
4. The Print dialog box will open. You can switch the orientation between Portrait and Landscape based on what's visually appropriate for your graph.
5. Make your selections and click **Print**.

## C. CHECKLIST FOR A PRESENTABLE GRAPH

The following items are required for a good graph (usually, content may vary).

1. Useful title
   a. More than just "Y vs. X"
   b. Not just "Concentration", but "Concentration of NaCl in..."
2. X and Y axis labels (**with units in parentheses**)
3. Appropriate numerical scale on x and y axes
   c. Conventional increments count by 1, 5, 10, 50 or 100, etc... (or 0.001, 0.005, 0.0025, etc...)
   d. Make sure all data are on the graph
   e. If a point has a y-value of 102 and the y axis only goes to 100, the point will not be visible
   f. Do not include extra decimals in the axis labels
   g. If the range is 0 to 200, the numbers should be in the format 0, 100, 200 and not in the format 100.000, 200.000, etc...
4. When printing, choose landscape or portrait depending on what looks best for the graph

Appendix 5

# Laboratory Equipment and Techniques

## A. COMMON LABORATORY EQUIPMENT

In any practical situation (like Chemistry lab), one of the first things you have to do is to examine the equipment available, to find out where the various pieces of equipment are stored, what they look like, and their names. Many of these items will be checked out to you in an individual locker or drawer, others will be readily available in common lockers or drawers in the laboratory, and still others will be put out into the laboratory only as needed. Figure One shows pictures of commonly-used pieces of laboratory apparatus made from glass, porcelain, plastic, or metal. As you check into lab during the first lab meeting (Appendix 1), be sure to familiarize yourself with these frequently-used pieces of apparatus.

## B. CLEANING AND DRYING LABORATORY EQUIPMENT

For most laboratory operations, it is of great importance that the equipment used (whether made of glass, metal, porcelain, or plastic) be clean. The general procedure is as follows. Wash the item well using warm tap water, soap, and a brush of appropriate size. Rinse it *thoroughly* with tap water to remove the soap and other material, then rinse once or twice with deionized water. (*Deionized water* is water from which dissolved materials in the form of charged particles (ions) have been removed.) Water will sheet and run evenly down clean glassware, but will form droplets on dirty glassware. Allow the item to drain thoroughly in an inverted position to dry if desired. It is relatively rare that dry glassware is absolutely necessary in the experiments in this manual.

If dry glassware is specified, however, the glassware should be thoroughly dried before use. Using a paper towel or kimwipe is fine. Sometimes, a material such as acetone is used to dry basic glassware, but it should *not* be used to dry clean *volumetric* glassware. (This liquid is toxic and flammable, however.) Acetone will mix completely with the water present, but will evaporate much more rapidly. Rinse the item with the acetone (it is *not* necessary to completely fill the piece being rinsed—doing so just wastes rinse liquid), pour the excess out into the appropriate waste container, and allow the item to dry in an inverted position. The compressed air line should not be used to dry glassware, since the compressed air sometimes contains water or other materials which could contaminate your cleaned glassware.

## C. DISPENSING AND DISPOSING OF CHEMICALS

It is very unlikely that you will be issued chemicals in prepackaged, pre-measured amounts. In any laboratory, there are bottles of liquid and solid chemicals on your lab bench or in some type

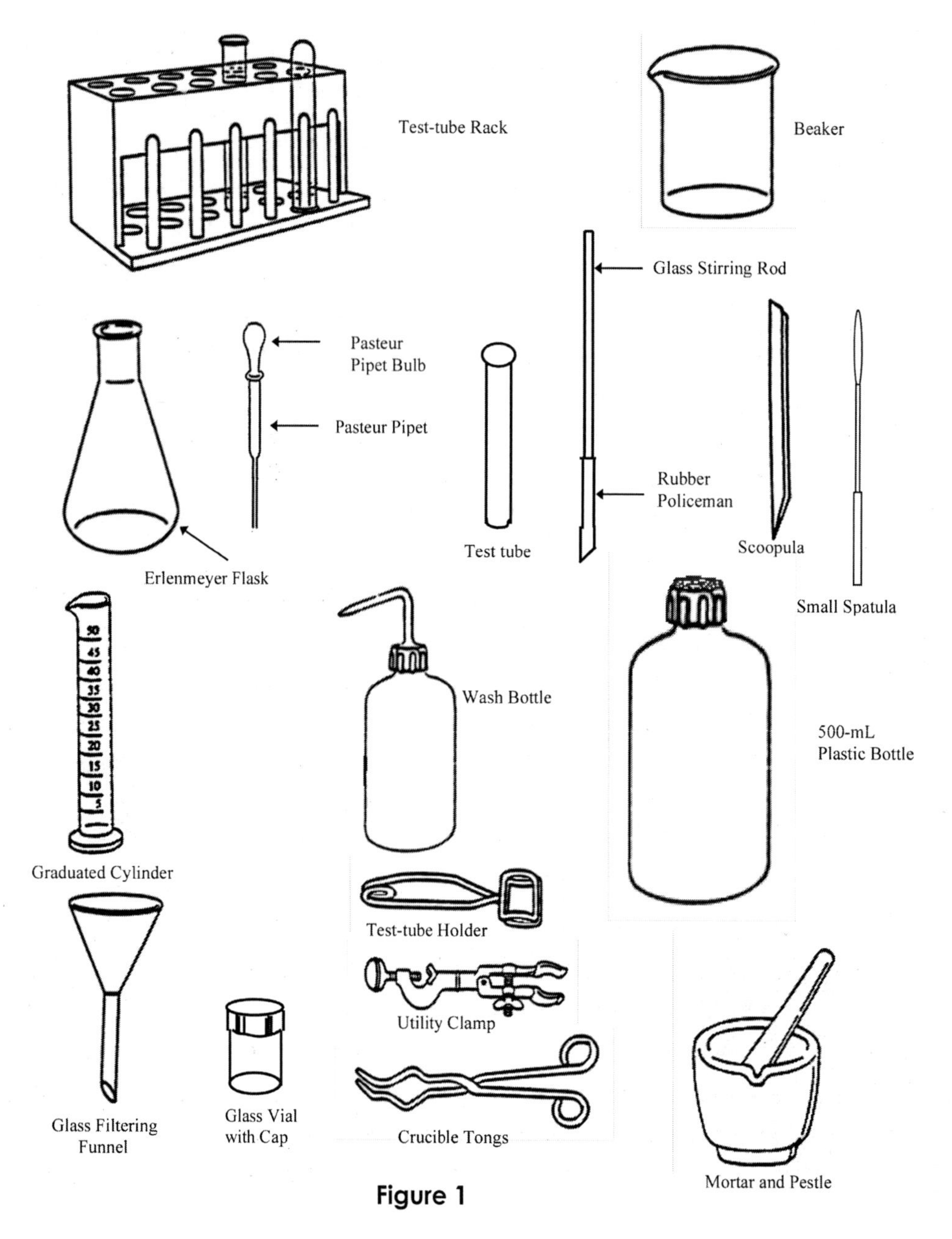


Figure 1

of storage area. As needed, small amounts of the appropriate chemicals are removed from these bottles for use. No matter what chemicals are used, what amount is needed, or what the form of the material (pure or in solution, liquid or solid), some general rules for the removal of chemicals from their storage bottles apply in all situations.

First, *no material, no matter how much is supplied, should be wasted.* Take *only* the amount you need at that time, or for that day's experiment, with perhaps a *slight* excess, but no more. If at a later time you need an additional amount, take it, but don't waste it. Once removed from the bottle, it *cannot* be put back due to the chance of contamination.

Second, *always be sure that the chemical withdrawn from the bottle is* ***exactly*** *the one you need. Read the label carefully* **before** you take anything from the bottle, then *read it again* ***before*** *you use* the material in the procedure specified in your instructions. Be sure that the name, formula, and concentration are exactly those specified in the procedure. This ensures that the procedure will be done both *correctly and safely.*

Third, ***nothing should be done to change the composition of the material in the stock bottle***. Once something is out of the stock bottle, it is out—*any excess should* ***NOT*** *be put back into the bottle.* Unless absolutely necessary, no implement (such as a spatula or dropper) should be placed into the bottle to remove the material. The material should be carefully poured from the bottle or jar into a container (beaker, weighing bottle, graduated cylinder, buret, etc.) without the aid of any implement if possible. The general rule is that *nothing* goes into a chemical bottle, (including excess reagent) and that only the chemical comes out. If at any time you do accidentally contaminate the stock bottle (or suspect that someone else has), take the bottle *immediately* to your instructor so that the situation can be dealt with in the proper manner. Do ***not*** just put the bottle back on the shelf.

Solid chemicals are generally supplied in wide-mouth bottles with either screw caps or ground-glass stoppers. To remove material from one of these bottles, remove the cap, set it down in an *inverted* position (to prevent contamination), and then pour he solid into the desired container. The rate of pouring can be controlled by "rolling" the bottle from side to side as you are pouring the solid. Liquid chemicals are usually supplied in narrow-mouth bottles. Dispensing liquids is done in basically the same manner as with solids. In this case, however, it is even more important to be careful when setting the stopper down on the lab bench.

Fourth, *any excess material should be disposed of in a safe, responsible manner. This also includes any material which you might have spilled onto the lab bench or outside of the bottle while transferring it.* This means that you should be concerned both with your own and your lab-mates' safety and with the safety of the community around you and your school. The general rule is *that materials should be disposed of in a manner that is consistent with their properties.* Materials that are insoluble in water should *not* be put down the drain—they should be placed in the appropriate container in the lab. Any toxic material should be placed in some sort of special container in the lab to be disposed of properly at a later time. Your instructor will inform you of the proper disposal container. Ordinary trash (e.g., paper towels) can be disposed of in a trash-can. Clean but broken glassware should be placed in the special containers designated for that purpose. In case of doubt, ask your instructor for the proper disposal technique for the specific material in question. More about proper waste disposal can be found in Appendices 2 and 3.

## D. MASS MEASUREMENTS

There are a number of different types of balances commonly used in Introductory Chemistry lab-oratories. In our laboratory, we use electronic top-loading balances, which have a single pan on the top of the balance for the object to be weighed and have a digital readout screen. This type of balance is very sensitive (it reads to the nearest 0.001 g), and must be treated with care in its operation. General rules for the use of the balance are as follows.

1. If you are assigned to use a particular balance, use *only* that balance throughout the semester. If it is not working properly, *tell your instructor*, who will either ensure that it is fixed or assign you to another balance. Whether or not you are assigned a balance, it is good practice to use the same balance throughout the course of a given day's lab work.

2. Always check to see that the balance is level. In a standing position, lean over and examine the leveling indicator located at the back of the balance. Do *not* lean on the bench on which the balance is sitting. The air bubble should be at the center of the circle. If it is not, notify your instructor *before proceeding*.

3. To operate properly, a balance and its surroundings must be kept clean. Solid chemicals spilled on the bench around the balance can generally be swept away with the brush and dust-pan provided for this purpose. If you spill anything, clean it up immediately; do *not* wait until the end of the lab period. Spills of liquids or sticky materials should be cleaned up only after consulting your instructor. *Any chemicals spilled directly on the balance must be reported to your instructor* ***immediately***. Do *not* attempt to clean it yourself.

4. Having checked that the balance pan is clean, check to make sure that the units of grams (g) are displayed. If not, have your instructor show you how to change them. Next, "zero" the balance by pressing the "zero" button. On most of our balances, this is the bar in front of the balance labeled as "→ O/T ←."

5. *Never* weigh any chemical directly on the balance pan. You should measure the mass of an empty, *clean*, *dry* container such as a weighing bottle, small beaker, or watch glass. *Remove the container from the balance pan before adding the sample to be weighed.* Reweigh the en-tire assembly. The mass of the sample is then equal to the total mass minus the mass of the empty container. *This is known as weighing by dif-*

*ference.* Corrosive samples, or those with a low boiling point should be weighed in a container with a stopper or lid of some type.

6. Objects to be weighed should be at *room temperature*. A hot object will give a mass reading that is lower than the correct value because of convection currents set up near the balance pan.

7. Record the mass (to ±0.001 g unless instructed otherwise) *directly* into your notebook. Do *not* write it on scraps of paper.

8. When you finish with your balance, be sure to leave it clean. Review step 3 above.

9. If at any time you experience difficulty in using your balance, *see your instructor*. Do not at-tempt to "fix" the problem yourself.

## E. MEASURING THE VOLUME OF A LIQUID

Many measuring devices exist for measuring volumes of liquids to different degrees of accuracy and precision. It is not always necessary to measure amounts with great accuracy. Sometimes, approximate amounts will do. For example, if you are told to "Dissolve the sample in approximately 30 mL of water," it is not necessary to use a buret to measure *exactly* 30.00 mL into the container. The instructions tell you that it is not necessary to do this, and although doing so would not be wrong, it would be a waste of your laboratory time. On the other hand, if the instructions say "Accurately measure 15.0 to 16.0 mL of the solution," this means that the volume should be somewhere in the range of 15 to 16 mL and should be measured and recorded accurately (at least to the nearest 0.1 mL), but that it does not have to be exactly 15.0, 16.0, or any other particular value. If the instructions read "Add 17.5 mL of solution," then the implication is clear—add 17.5 mL, no more, no less, and measure accurately to the nearest 0.1 mL or better.

Approximate volumes can be measured in a number of ways. Most flasks and beakers have approximate graduations that are reliable to ±5 mL; this is adequate for many labora-tory opera-tions. Volumes can also be roughly estimated by "calibrating" small test tubes or by counting drops dispensed from a dropper. (Typically, 20 to 25 drops equal 1 mL, but for droppers with very fine tips, 40 or more drops may equal 1 mL. You should check this for yourself early in the semester.)

**Reading a Meniscus.** Whenever water or an aqueous solution is placed in a narrow vertical glass tube (such as a graduated cylinder, pipet, or buret), the surface of the liquid will not be flat. Because of the attraction of the glass for the water, the part of the liquid nearest the glass will be pulled upward, and the center of the liquid surface (farthest from the glass walls) will be lower than the outer part. This curved surface is called a *meniscus*. In large-diameter glassware such as a beaker or flask, this is of little importance, but in a narrow calibrated tube such as a graduated cylinder, pipet, or buret, it must be considered. A close-up view of the liquid in a cylinder is shown below in Figure Two (a). Where on this curved surface should the volume be read? It is obvious that an error of as much as 0.2 mL could be introduced by reading the volume at the incorrect place. Volumetric glassware is always calibrated so that the correct reading will be obtained by reading the *bottom* (i.e., the lowest point) of the meniscus—that is, at the *center* of the liquid surface, *not* at the walls of the container. The correct volume reading in Figure Two (a) is 15.6 mL. Notice that the reading was not just made to the nearest graduation marking, at 16 mL. It is common practice to read and record volumes to one estimated figure beyond the calibration markings.

What would be the correct volume reading for the buret pictured in Figure Two (b)? At first glance, you might say that the reading was 37.62 mL, but a little more thought would lead to the conclusion that this reading *cannot* be greater than 37 mL, since the bottom of the meniscus is between the 36- and 37-mL marks. You will notice that the markings shown in the figure seem to be "upside down"—that is, they are larger toward the bottom of the tube rather than toward the top. It is not uncommon for burets and for some pipets to be calibrated in this manner, since the amount *removed* or *delivered* is often more important than the amount *contained* in the vessel. You must *always* check, before making your reading, how the instrument is calibrated—in a "normal" fashion (as in Figure Two (a), or "upside down" (as in Figure Two (b)). In this case, the correct volume reading in Figure Two (b) is 36.38 mL. Note again that the reading is taken to one figure beyond the smallest scale division, not just to the nearest mark.

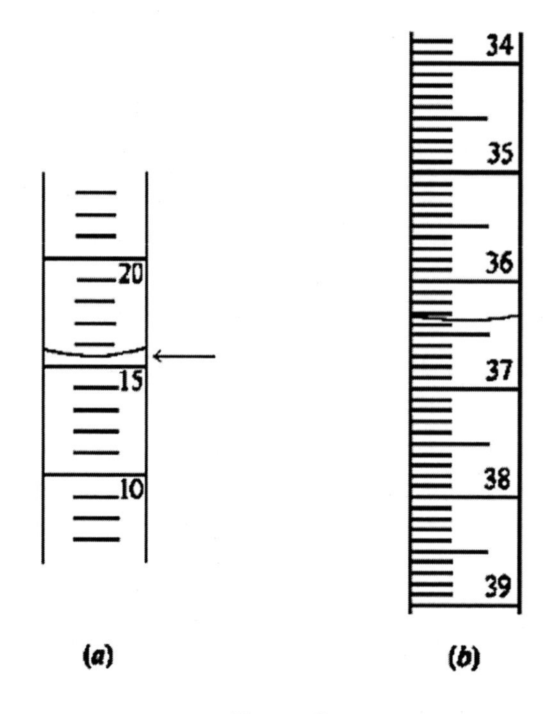


Figure 2

Lastly: To avoid parallax error in your readings, the bottom of the meniscus should always be *at eye level* when the reading is being taken.

**Graduated Cylinders.** For volume measurements to the nearest 1 mL or (at best) 0.1 mL, graduated cylinders are normally used. These are tall glass or plastic cylinders with a wide base for stability. They are commonly found in sizes of 10, 25, 50, and 100 mL in Introductory Chemistry laboratories, but are available in sizes up to 5 L for other purposes. These can be read quickly and with reasonable accuracy. They are usually calibrated to *contain* a certain volume of liquid, and would be marked "**TC**" to indicate this. Glassware marked in this way will contain the specified amount, but will transfer or deliver less than this when the contents are poured out. This is because a small amount of liquid will remain on the container walls after most of the liquid has flowed out. If you must *deliver* a precise amount of liquid, then a graduated cylinder is not reliable, and a pipet or buret (discussed in the next section) must be used. To read a graduated cylinder to the nearest mL, it is not usually necessary to read the exact bottom of the meniscus; to obtain precision of ± 0.1 mL, the bottom of the meniscus must be read. Generally, only two (or at most three) significant figures are obtainable with a graduated cylinder, even in the best of circumstances. The smaller the graduated cylinder, the more precisely (i.e., with more significant figures) it can be read.

**Pipets.** Volume measurements more precise than ± 0.1 mL must be made using a pipet or buret. A *pipet* is a narrow glass or plastic tube tapering to a fine point at one end and having at least one calibration marking on it. Two types of pipet are commonly encountered in Introductory Chemistry laboratories: the *volumetric* pipet and the *measuring*, or *Mohr*, pipet. A volumetric pipet (Figure Three (a)) often has a bulge in the center (as shown) and has only one calibration marking on the upper part of the tube. The measuring pipet (Figure Three (b)) is a constant-diameter tube with markings along most of its length. Measuring pipets are made in sizes ranging from 1 mL to 50 mL, but are most commonly found in a 5- or 10-mL size. Volumetric pipets can be found in sizes ranging from 0.1 mL to 100 mL, but most commonly in 5-, 10-, or 25-mL sizes.

The two types of pipets are used in basically the same manner. After checking to see that neither the tip nor the top opening is chipped (if it is, bring it to your instructor), the pipet is cleaned by rinsing with an appropriate liquid, so that the liquid drains cleanly from the tip and forms no beads on the inner walls. Then, using a pipet bulb or pump, the liquid to be measured is drawn

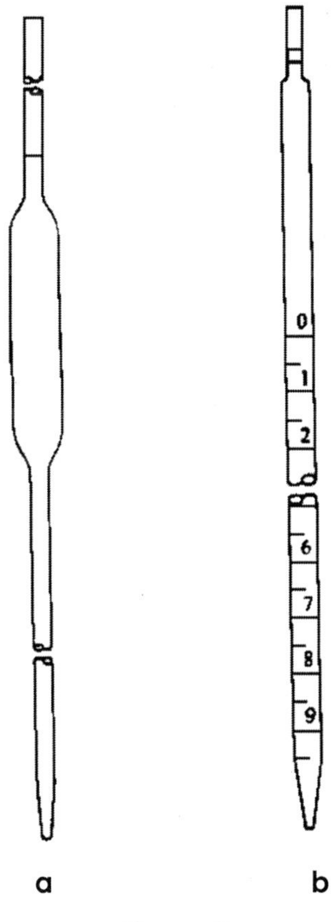

Figure 3

into the pipet to a point above the top calibration marking. The liquid is drained out to the mark, then the pipet is drained into the receiver vessel. It cannot be emphasized too strongly that **it is absolutely necessary to use a pipet bulb or pump to suck the liquid into the pipet. It is an extremely unsafe practice to use mouth suction for this.** Many of the liquids that might be pipetted in a general chemistry laboratory are toxic, foul-tasting, irritating, or otherwise unplea-sant. Use of a pipet bulb or pump eliminates the possibility that any liquid (unpleasant or other-wise) can be accidentally sucked into the mouth and/or ingested.

To perform the above set of operations using a pipet bulb, the sequence shown in Figure Four should be followed. The top end of the pipet should be moistened slightly, the bulb squeezed to expel air (Figure Four (a)), the bulb placed in contact with the pipet, and the liquid drawn into the pipet The bulb should *not* be forced totally over the end of the pipet so that the bulb could be used as a handle. At all times, the pipet should be held in one hand, and the bulb in the other. Once the liquid has been drawn into the pipet above the desired calibration mark (but *not* into the bulb!) (Figure Four (b)), the bulb is removed and the end of the pipet is quickly covered with a moistened fingertip (Figure Four (c)) (*not* a thumb, which gives less control over the liquid level), as shown in Figure Four (d). The liquid level is carefully allowed to fall until it is at the calibration mark (Figure Four (e)) (remember to read the bottom of the meniscus; drain the excess liquid into a waste container), the outside of the pipet is dried, and the liquid is then drained into the receiver vessel, as shown in Figure Four (f). The pipet should be held *vertically* when readings are taken and when it is being drained. While draining, the tip of the pipet should be held against the inside wall of the receiver vessel.

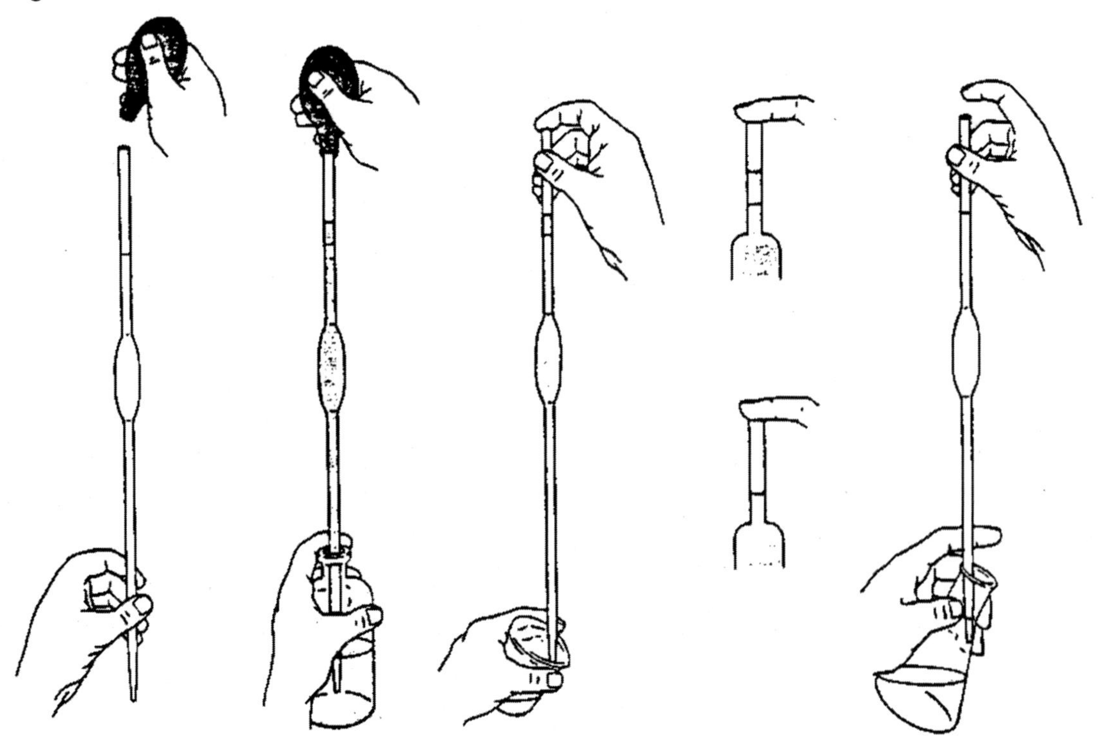

Figure 4

If using a pipet *pump*, the procedure is essentially the same as that described above for a pipet bulb. In this case, the plunger on the pump should be fully depressed (analogous to squeezing the pipet bulb) before the pump is connected to the pipet. Then, the knurled wheel is turned to draw the liquid into the pipet (analogous to releasing the bulb) above the calibration mark. Then the pump is carefully *removed from the pipet* (it is a pipet "pump," not a pipet "handle"), the opening is covered with a forefinger, and the process is completed as described above.

A volumetric pipet is calibrated to *deliver* (as shown by its "**TD**" marking) exactly one specified amount—10.00 mL, 25.00 mL, and so on. It should be allowed to drain freely until no more liquid comes out. At this point, a small amount of liquid will still be in the pipet. This "extra" volume is *beyond* the volume stated and *should **not** be blown or rinsed out*. Removal of *all* the liquid in this type of pipet will result in the transfer of *more* than the stated volume by some unknown amount. As indicated earlier, volumetric pipets can be used to give volumes reliable to two decimal places *if used properly*.

A measuring pipet can be used to transfer (deliver) any volume between 0 mL and the maximum volume of the pipet. In this case, it is necessary to take *two* volume readings—the initial reading and the final reading after the transfer of the liquid. (This is similar to using a buret, described below.) The amount delivered is then the *difference* between the two readings. This type of pipet should never be drained below the lower calibration mark if an accurate volume is needed. In reading this type of pipet, be sure to note whether it is calibrated "normally" or "upside-down." The general use, filling, and so forth for a measuring pipet are identical to that for a volumetric pipet, as shown in Figure Four and described above. Depending on their size, measuring pipets can normally be read to the nearest 0.01 mL, giving three or four significant figures as a result.

One last point concerning the use of pipet bulbs and pipet pumps: If at any time, *any* liquid (including pure water) is sucked into the bulb or pump, the bulb or pump must be thoroughly drained, rinsed, and dried before being used again. If liquid is left in the bulb or pump, it is possible that some can accidentally flow into the pipet, thereby contaminating (or at least diluting) the solution being transferred. If you accidentally get *any* liquid into a pipet bulb or pump (or if you pick a wet bulb or pump from the supply), be sure to put it aside and *give it to your instructor* for proper cleaning.

**Burets.** Pipets are useful for delivering "round" volumes such as 5.00 mL. When repeated mea-surements of non-round volumes such as 19.57 mL are needed, a buret is the most common choice. Burets are long graduated glass tubes, available in many sizes from 1 to 100 mL. Most commonly, 50-mL burets are used in general chemistry laboratories. At the bottom a buret has a glass or plastic stopcock for controlling the flow rate of the liquid. A typical setup for a buret is shown in Figure Five. The buret is clamped to a ringstand in a *vertical* position by means of a special type of clamp called a buret clamp.

Before making any measurements, the buret must be thoroughly cleaned (so that no water droplets adhere to the inner walls), then rinsed with the solution that is to be measured. This is done by pouring 5 to 10 mL of the solution into the buret (with the stopcock *closed*!), then tilting the buret to an almost horizontal position and rotating it so that the entire inner surface comes in contact with the liquid. The wash liquid is then allowed to flow out *through the tip* to rinse that section as well. The rinse liquid is discarded into the sink or other appropriate waste container. The rinsing process is then repeated once or twice more. Once the buret is clamped back ontothe ringstand, it is filled to *above* the top calibration mark, then drained to get the meniscus down to the calibrated portion of the buret. All air bubbles should be removed from the tip of the buret and any liquid on the outside of the buret should be dried off. If a funnel was used to aid in filling the buret, it should be removed. (Liquid can be temporarily trapped between the funnel and the buret wall. If this liquid flows out during the measurement, the readings will be in error.) The initial volume reading is then accurately made. Most burets are calibrated "upside down," so be sure that your reading is correct.

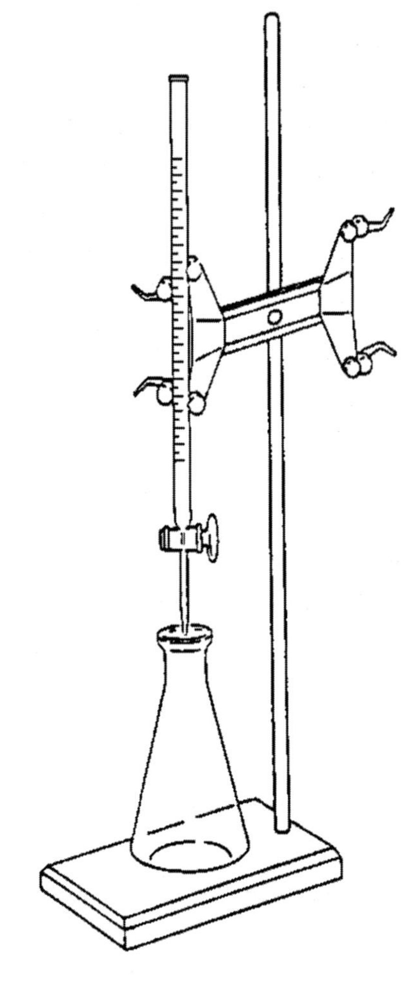

Figure 5

The required amount is then allowed to drain from the buret by opening the stopcock. After closing the stopcock, the final reading is taken. As with a measuring pipet, neither the initial nor the final reading has any meaning by itself. The meaningful number is the *difference* between the initial and final readings, since this represents the volume *delivered* by the buret. For example, if the initial reading were 5.72 mL, this would not mean that the buret contains 5.72 mL of liquid. In fact (assuming that a 50-mL buret were being used), more than 45 mL of liquid would be actually contained in the buret. If liquid were removed so that the final reading were 37.92 mL, the volume delivered would be 37.92 – 5.72, or 32.20 mL.

**Volumetric Flasks.** When you are instructed to dilute a sample to a specified volume such as 50.0 mL, a *volumetric flask* is used. A typical volumetric flask is shown in Figure Six below. They are available in volumes ranging from 1 mL to 2 L or more, but sizes between 25 and 500 mL are those most commonly used in the Introductory Chemistry laboratory. They have a single mark on the neck, and are calibrated to *contain* the specified volume when the bottom of the meniscus is at that mark. They are precise to at least ± 0.1 mL, and generally are precise to ± 0.01 mL, depending on the size and the manufacturer. Their closure may be a glass or plastic stopper, or a plastic snap-cap.

To use a volumetric flask, the sample to be dissolved or diluted (either solid or liquid) is placed in the cleaned flask, solvent is added until the flask is about 3/4 full, and the flask is swirled to mix/dissolve the contents thoroughly. Then, additional solvent is added until the bottom of the meniscus is *exactly* at the mark on the neck. The stopper or cap is then put in place, and the flask is inverted (while holding the closure tightly) to mix the solution. Allow the bubble of air from the neck of the flask to rise fully to the bottom of the flask (now at the top), then return the flask to its normal orientation. Again allow the air bubble to rise fully to help mix the solution. Repeat this process a few times to ensure thorough mixing of the solution. Do not rush the mix-ing operation—let the bubble of air do the mixing to obtain the best results.

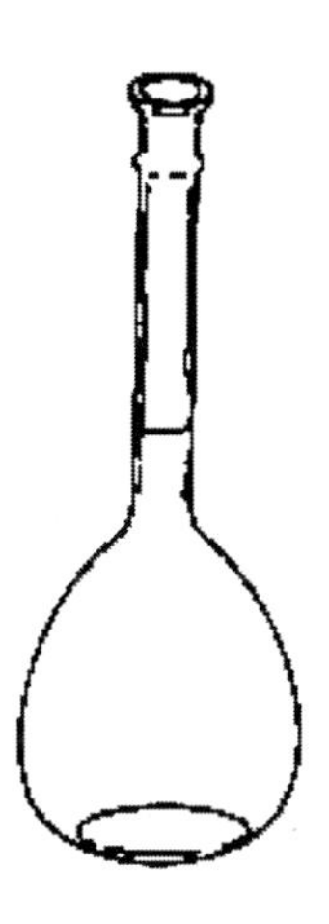

Figure 6

One of the most crucial parts of this operation is filling the flask to the mark. If you add liquid to beyond the mark, the only way to complete the solution preparation without error is to empty the flask, clean it thoroughly, and measure a new sample to be diluted. This obviously takes a lot of extra laboratory time, and wastes material as well. Leaving the excess liquid in the flask introduces an error in the volume reading, and removing sufficient liquid to bring the level back down to the mark runs the risk of removing some of the dissolved sample as well as the solvent, introducing a different error. It is therefore important to be extremely careful in diluting the solution *to, but not above, the mark*. This final part of this operation should be done with a dropper pipette.

## F. SEPARATING A MIXTURE OF LIQUID AND SOLID

Many times, either in analyzing a sample or in the preparation of a material, a mixture of solid and liquid is obtained. If the solid settles readily, and if an incomplete separation of liquid and solid is acceptable, decantation can be used to separate the components. In this procedure, the solid is simply allowed to settle and the liquid is carefully poured from the top of the sample. Usually, some of the solid is also transferred out with the liquid, resulting in an incomplete separation and recovery of the solid. The separation can be made more quantitative by first centrifuging the mixture (see Appendix 6). If a truly complete separation and recovery of the solid is required, however, filtration must be used. In this procedure, the mixture is poured through a porous filter paper, which traps the particles of solid and allows the liquid to pass through.

### 1. Gravity Filtration

The setup for a gravity filtration is shown in Figure Seven. The funnel is supported by an iron ring or a funnel stand, and the tip of the funnel is placed in a vessel to catch the liquid.

A circular piece of filter paper is folded into quarters as shown in Figure Eight. The diameter of the paper used should be approximately 1 cm *less* than twice the diameter of the top of the funnel. The filter paper is first folded exactly in half (Figure Eight (a)), then the second fold is made so that one corner of the paper is about 3 mm (1/8 in.) inside the other (Figure Eight (b)). The paper is then opened into a cone (Figure Eight (c)) so that the shorter edge of the paper will be pressed against the funnel wall and be covered by the longer edge of the paper (Figure Eight (d)).

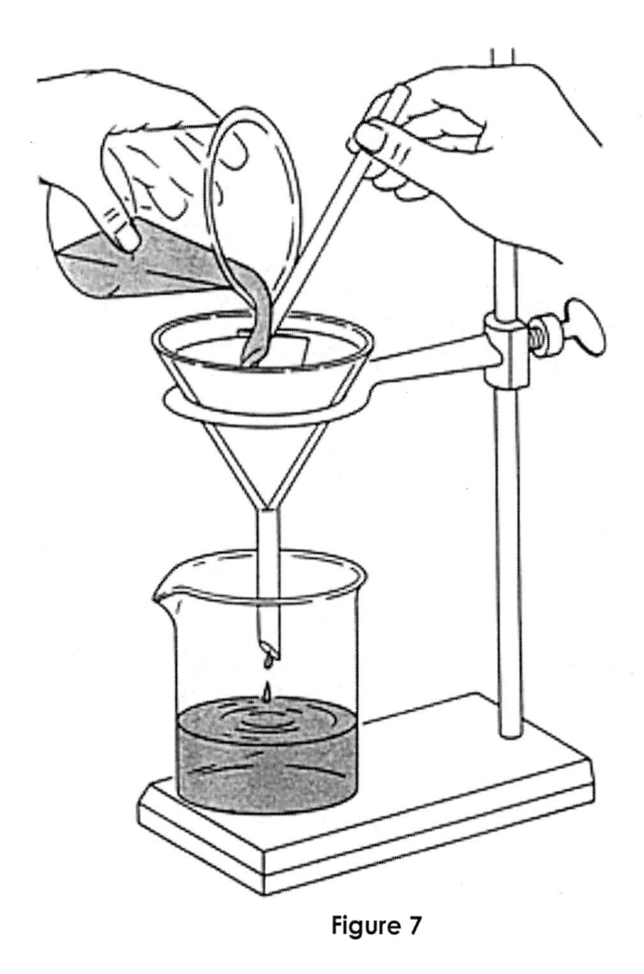
Figure 7

The paper is thoroughly wet so that it adheres to the funnel and the liquid forces out all the air bubbles from the funnel stem. A full column of water in the stem will create a gentle suction that speeds the filtration.

The funnel is returned to the iron ring, the mixture to be filtered is gently swirled or stirred, and then carefully poured into the funnel. Care must be taken not to fill the *paper* more than two-thirds to three-fourths full. It is usually a good idea to guide the mixture into the funnel by using a glass stirring rod, as shown in Figure Seven. As liquid drains out, more of the mixture is added to the funnel until all the mixture has been transferred. Continue to swirl or stir the mixture to keep the solid suspended and enable you to transfer most of it into the funnel. If any solid remains in the original vessel, water or the *filtrate* (the liquid that passed through the filter) is used to wash the solid into the funnel. When all the liquid has drained from the filter, the paper is removed from the funnel and treated as instructed in the procedure or by the instructor. Be sure to do this carefully—wet filter paper tears very easily.

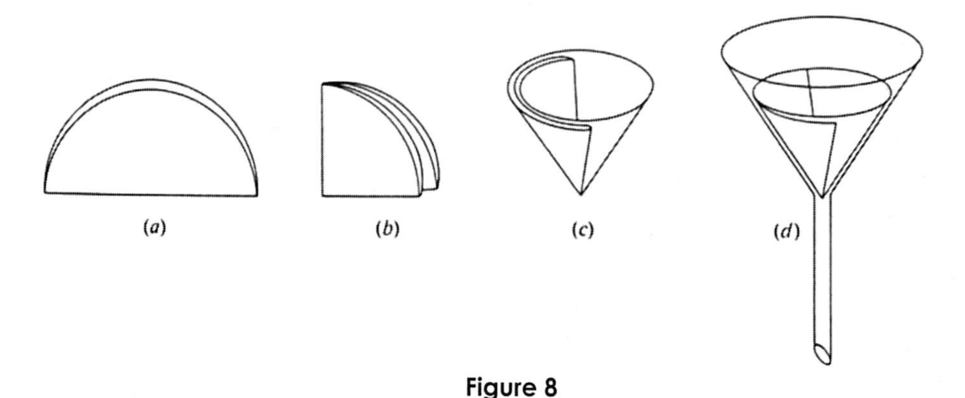

Figure 8

## 2. Suction Filtration

Gravity filtrations are reasonably easy to set up and generally give fairly complete separation of the solid and liquid. Their major disadvantage is that they tend to be slow. To speed the filtration, a vacuum or a *suction filtration* is often done. These are faster and allow faster drying of the solid, but sometimes do not give as complete a separation of the solid and liquid, as does a gravity filtration. The set-up for a suction filtration is shown in Figure Nine. The funnel for this type of filtration is made of plastic, and is made in two pieces. This type of funnel is called a *Büchner funnel*. The flat plate in the center of the funnel supports a circle of filter paper. The paper should be chosen so that it *exactly* fits in the funnel–it should not have to be cut or folded, and it should cover all the holes in the plate. The funnel is placed in a rubber collar or sleeve, or in a rubber stopper. This ensures that the seal between the funnel and the filter flask will be vacuum tight.

The filter flask looks like a typical Erlenmeyer flask, except that it is made of very thick glass (to help prevent breakage under vacuum) and it has a glass sidearm near the top. The entire apparatus is connected to a source of vacuum using heavy-walled rubber or plastic tubing. The vacuum source is a vacuum outlet in the lab (similar to a gas jet). Often, a trap is placed between the source of vacuum and the system used for the actual filtration. This helps prevent the backup of unwanted material into the system. *Most commonly, the filter flask is clamped to a ringstand to prevent its tipping over and spilling its contents onto the laboratory bench.*

Thoroughly clean the Büchner funnel inside and out, rinse well with deionized water, and shake out the excess water. The filtration is performed by assembling the apparatus as shown, including a piece of filter paper of appropriate size. Turn on the suction, and moisten the filter paper with a *small* amount of water (to make sure that it seals tightly to the perforated plate). Gently swirl or stir the mixture to be filtered to aid in loosening the solid on the bottom and sides of the container, resulting in an easier and more complete transfer of all of the solid into the funnel. Carefully pour the mixture to be filtered onto the paper. As the suction pulls liquid through the funnel, more of the mixture can be added to the filter paper.

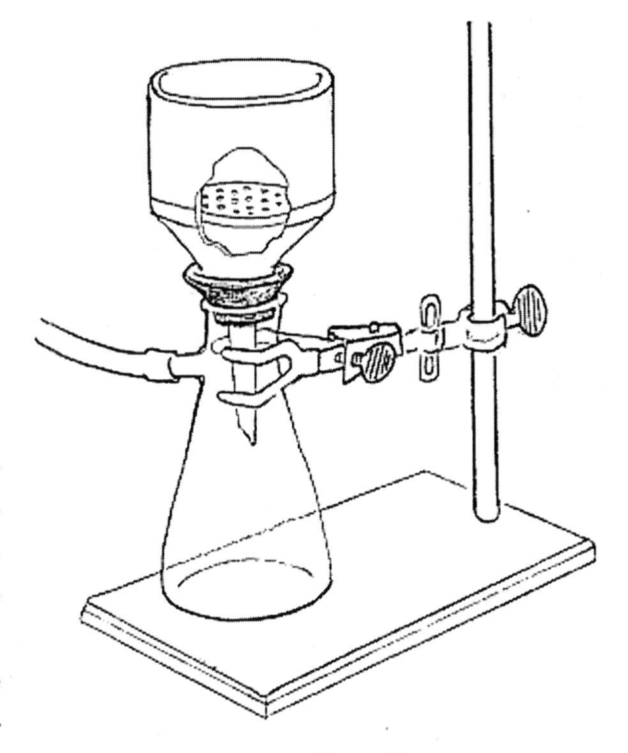

**Figure 9**

Once all of the mixture has been added to the funnel, any solid remaining in the original container should be rinsed into the funnel. This can be done using the *filtrate* (the liquid which has passed through the filter), or with the first sample of the liquid used to wash the solid filtered from the mixture.

Once the liquid has been pulled through the solid, the solid should be washed with some appropriate wash liquid. To do this, turn the vacuum off, pour the wash liquid onto the solid, making sure that the *entire* sample is contacted by the wash liquid, then turn the vacuum back on to pull the wash liquid through the solid. While the wash liquid is on the solid, you can stir the solid *gently* with a glass rod or metal spatula to be sure that the entire sample is washed thoroughly. But: Be careful—wet filter paper tears very easily. Repeat the washing if directed. The vacuum can be left on for a few minutes to aid in the drying of the solid. If the filtrate is poured from the filter flask first, the drying process is speeded up.

# Appendix 6

# Microchemical Techniques

## A. Adding and Mixing Reagents; Forming a Precipitate

The techniques described here are used when small amounts (less than about 1 mL of solution) of reagents are being used. Since you are using such small quantities of reagents, and some of the tests you will be performing are very sensitive to impurities, you must be very careful not to cross contaminate the solutions as you are dispensing them. Keep all your glassware clean and rinsed thoroughly with deionized water. The reagent bottles are fitted with dropper caps, so there is no need to use a separate dropper to add a test reagent to another mixture. Be sure that the dropper portion of the bottle does not come in contact with the mixture in the test tube.

Once the specified amount of test reagent is added, the mixture must be *thoroughly* mixed. Do **NOT** simply place your finger or a cork over the mouth of the tube and shake! Even though this gives effective mixing, it can result in contamination of the sample and injury from contact with the contents of the tube. If the tube is less than half full, sideways shaking of the tube works well. Either hold the tube near the mouth and shake *sharply* sideways a few times, or hold the tube near the mouth with one hand and tap the bottom of the tube *sharply* with the other. Gentle shaking or tapping won't work as well. This technique may not work well if the tube is more than half full. In that case, pouring the mixture back and forth a few times between two clean tubes is probably the best method.

Using a stirring rod or spatula is the method often chosen by students, but is actually one of the *least* effective methods of mixing the contents of a test tube. If you do choose to use a stirring implement, be sure that it is clean before stirring the mixture, and use a glass stirring rod, rather than a metal spatula, which could react with the mixture being stirred. Also be sure to use a combination of up-and-down and circular motions to ensure complete mixing of the contents of the tube, both top-to-bottom and side-to-side.

## B. Using a Centrifuge

If the mixing of two or more solutions produces an insoluble material (i.e., a *precipitate*), you often will be told to isolate the precipitate to perform further tests. With small amounts like you will be using, filtration of the mixture is both cumbersome and time consuming, and has the potential to waste or lose a significant amount of sample, so is rarely if ever done. The most common way of separating a small amount of precipitate from a small volume of solution is to use a centrifuge. This compresses the solid at the bottom of the tube so that it can more easily be separated from the supernatant solution. A typical *centrifuge* is shown in Figure One below.

If you are centrifuging more than one mixture at a time, you should label the tubes before placing them in the centrifuge, and you should place the tubes in a symmetrical arrangement so that they are *balanced*. This would mean using holes 1 and 4 for two tubes; holes 1, 3, and 5 for three

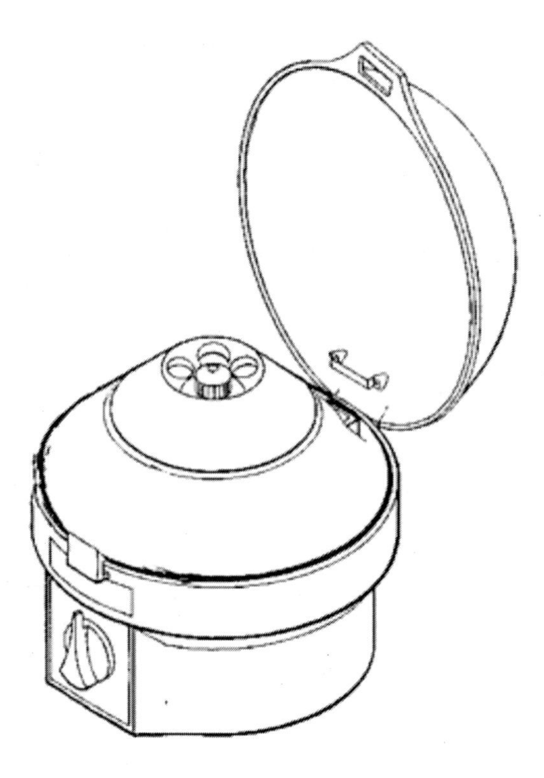

**Figure 1**

tubes, etc. (See the Figure in Experiment 6.) If the centrifuge is not balanced, it can wobble, causing damage to the apparatus and possibly causing it to “walk” off the lab bench. The centrifuge should be *closed* before turning it on. Once you turn it on, allow it to reach its maximum rotation rate and let it spin at this rate for at least 15 s to ensure complete compaction of the precipitate at the bottom of the tube. When you turn it off, *allow it to slow to a complete stop* ***on its own***. Do ***not*** slow or stop the centrifuge with your hand! Doing this can result in a less-than-completely separated mixture (since sudden stopping can re-mix the mixture by jarring it) and also in personal injury (from friction burns to the hand and/or possibly cuts from broken glass).

Remove the tube(s) from the centrifuge and examine it. The solid should be completely compacted at the bottom of the tube and the supernatant liquid should be clear (not cloudy). If the precipitate has not been completely compacted, return it to the centrifuge for another period of spinning, until complete separation is attained. Failure to follow *all* these steps *completely* can result in incomplete separation of the mixture, damage to the centrifuge, and/or personal injury.

Once the mixture is well-separated, and the precipitate is completely compacted at the bottom of the tube, the supernatant liquid can simple be poured from the solid (*decanted*). This doesn't give a perfect separation of solid and liquid, but is more than adequate for most purposes.

## C. Washing a Precipitate

As noted above, centrifugation followed by decantation of the supernatant doesn't give a perfect separation of the solid and liquid. Sometimes, a small amount of the solid is transferred out when the liquid is poured from the tube, but this is normally negligible if the centrifugation has been performed well. However, there is invariably some of the liquid trapped with the solid, and which is not poured from the tube. This liquid will contain dissolved materials which may interfere with succeeding tests performed on the precipitate. To have a "clean" sample of solid for further testing, the solid must be thoroughly washed.

To do this, add the specified amount of the wash liquid (usually, but not always, deionized water) to the solid, mix thoroughly with a glass stirring rod (being sure to break up any clumps of solid to ensure thorough contact with the wash liquid), and then re-centrifuge the mixture as described in Part B above, discarding the wash liquid into the appropriate waste container. This washing process will remove adhering liquid (with any dissolved materials) which might have been trapped in the solid. If instructed to do so, perform a second washing in the same manner.

## D. Use of Test Papers

Test papers are a fast, convenient way to detect or "measure" some component of a solution (or gas). These are pieces of paper (essentially filter paper) which have been impregnated with a solution of an indicator material and then dried. One example is lead acetate paper, which darkens (due to the formation of brown/black PbS) when moistened and exposed to hydrogen sulfide ($H_2S$) gas, or is wet with a solution containing $S^{2-}$ ions. Another example is *litmus paper*, which contains the acid-base indicator *litmus*. This material is blue in base (pH above 8), red/pink in acid (pH below 6), and purple if the solution is close to neutral (pH between 6 and 8). Therefore, if red litmus paper is moistened with a basic solution, it turns blue, and if blue litmus paper is moistened with an acidic solution, it turns pink. Another common form of test paper is *pH paper*, which has been impregnated with an acid-base indicator (or, more commonly, a mixture of two or more indicators) which changes color over a specified pH range. There are many var-ieties available, depending on the pH range over which they change color. *Wide-range* pH paper undergoes a series of color changes as the pH is varied over the entire 0-14 pH range. *Short-range* pH paper undergoes a color change over a range of about 2-4 pH units, and is available in a dozen or more pH ranges.

To test a *solution*, a test paper is used in the following way: A glass stirring rod is immersed in the solution and then withdrawn, removing a small amount of the solution on the tip of the rod. The rod is then touched to the test paper and the color is noted. You should note the color of the paper immediately after touching it with the solution. This is because the color may change as the mixture is exposed to the atmosphere. If the solution itself is colored, you should note the color of the *outermost* part (i.e., the leading edge) of the moistened spot; that will give the color least affected by the color of the solution. Do **not** put the test paper directly into the solution being tested. Doing this wastes solution (of which you may have only a small volume to begin with) and may introduce paper fibers and/or indicator dye which could contaminate the solution.

To test a gas, moisten the test paper with deionized water, expose it to the gas, and then immediately note the color of the paper. As with a solution, the color may change as the mixture is exposed to the atmosphere.

If you are given an instruction such as “acidify the solution with 6 M HCl,” or “add sufficient reagent to neutralize the previously-added reagent,” you should first test the pH of the solution (with litmus or pH paper) so that you know how far from neutral you are starting. Then add a small amount of the adjusting solution (acid or base), mix the solution *thoroughly*, and test it as outlined above. Continue this operation until the desired pH is attained. The amount of adjusting solution added in each increment is a function both of the volume of the solution and how far away from the desired pH the solution is. The smaller the volume of the solution being adjusted and the closer to the desired “endpoint” you are, the smaller should be the increments of the adjusting solution. Be careful that the liquid withdrawn on the glass rod is the solution being tested, not traces of some (possibly different) liquid from the walls of the container. (Although, if you have indeed mixed the solution thoroughly, this problem should not occur.)

# Appendix 7

# Some Common Polyatomic Ions

Common Monatomic Ions

| Cations | | | Anions | | |
|---|---|---|---|---|---|
| **Charge** | **Formula** | **Name** | **Charge** | **Formula** | **Name** |
| 1+ | $H^+$ | hydrogen | 1- | $H^-$ | hydride |
| | $Li^+$ | lithium | | $F^-$ | fluoride |
| | $Na^+$ | sodium | | $Cl^-$ | chloride |
| | $K^+$ | potassium | | $Br^-$ | bromide |
| | $Cs^+$ | cesium | | $I^-$ | iodide |
| | $Ag^+$ | silver | 2- | $O^{2-}$ | oxide |
| 2+ | $Mg^{2+}$ | magnesium | | $S^{2-}$ | sulfide |
| | $Ca^{2+}$ | calcium | 3- | $N^{3-}$ | nitride |
| | $Sr^{2+}$ | strontium | | | |
| | $Ba^{2+}$ | barium | | | |
| | $Zn^{2+}$ | zinc | | | |
| | $Cd^{2+}$ | cadmium | | | |
| | $Hg^{2+}$ | mercury (II) | | | |
| 3+ | $Al^{3+}$ | aluminum | | | |

Common Polyatomic Ions

| Charge | Formula | Name | Charge | Formula | Name |
|---|---|---|---|---|---|
| 1- | $CH_3COO^-$ | acetate | 2- | $CO_3^{2-}$ | carbonate |
| | $NO_3^-$ | nitrate | | $CrO_4^{2-}$ | chromate |
| | $NO_2^-$ | nitrite | | $Cr_2O_7^{2-}$ | dichromate |
| | $ClO_4^-$ | perchlorate | | $SO_4^{2-}$ | sulfate |
| | $ClO_3^-$ | chlorate | | $SO_3^{2-}$ | sulfite |
| | $ClO_2^-$ | chlorite | | $O_2^{2-}$ | peroxide |
| | $ClO^-$ | hypochlorite | 3- | $PO_4^{3-}$ | phosphate |
| | $HCO_3^-$ | bicarbonate or hydrogen carbonate | | | |
| | $MnO_4^-$ | permanganate | 1+ | $H_3O^+$ | hydronium |
| | $CN^-$ | cyanide | | $NH_4^+$ | ammonium |
| | $OH^-$ | hydroxide | 2+ | $Hg_2^{2+}$ | mercury (I) |

# Appendix 8

# General Solubility Rules of Ionic Compounds in Water

**Soluble**

1. All salts of Group IA (alkali metals) are soluble.

2. All common compounds of ammonium are soluble.

3. All salts of acetates, chlorates, perchlorates and nitrates are soluble.

4. All salts of halides ($Cl^-$, $Br^-$ and $I^-$) are soluble except those of silver(I), copper(I), lead(II), and mercury(I).

5. All salts of sulfate are soluble except those of barium, calcium, lead(II), silver and strontium.

**Insoluble**

1. All salts of carbonate, phosphate and sulfite are insoluble, except those of group IA (alkali metals) and ammonium.

2. All oxides and hydroxides are insoluble except those of group IA (alkali metals), calcium, strontium and barium.

3. All salts of sulfides are insoluble except those of Group IA(alkali metals) and IIA (alkaline earth metals) and of ammonium.

# Appendix 9

# Activity Series of Common Metals

Metals are listed according to their ability to displace hydrogen from an acid (or water) to form H2. The most reactive metals are listed at the top. These are the metals that give up electrons most readily and are considered to be the *strongest reducing agents*. The least reactive metals are at the bottom and are considered to be the *weakest reducing agents*. Any elements higher in the activity series will reduce the ion of any element lower in the activity series.

## Activity Series of Common Metals

**Oxidation Reaction**

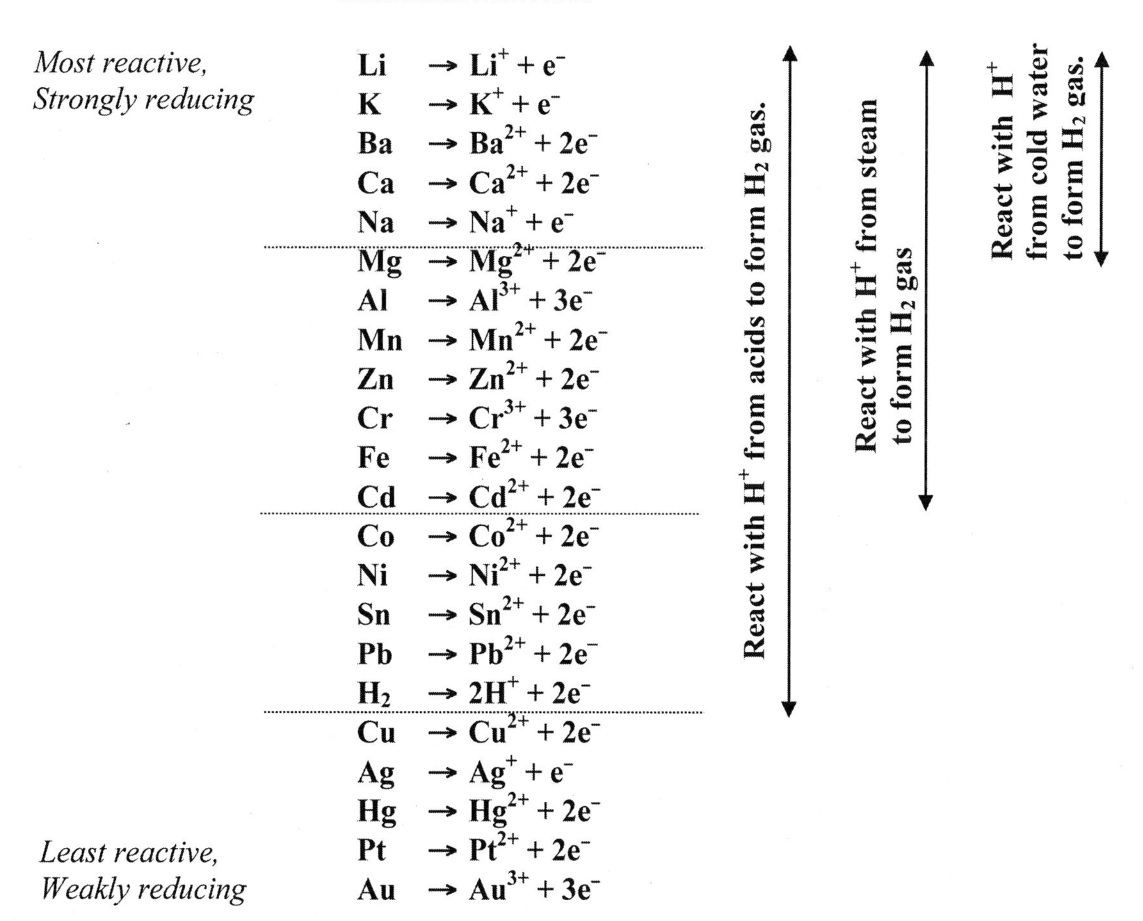

## Appendix 10

# Instructions for Molecular Model Kits

### "Ball & Stick" Model Kit

The "Ball & Stick" model kit consists of wooden balls of various colors, representing atoms. Each *unpainted* wooden stick represents a *bonding pair* of electrons (i.e. one covalent bond). Each *painted* stick glued into the ball represents a *nonbonding pair* of electrons (also called a *lone pair*). Do **not** attempt to remove any of the painted sticks from the balls.

The number of holes in a ball represents the number of bonds the atom can have. Since the maximum number of holes in the balls found in this model kit is four, you can only build models for atoms that obey the "Octet Rule." In other words, you cannot build models of atoms with an expanded octet with this type of model kit.

Examine the Lewis structure for the species whose model you wish to build. Choose a ball for each atom by selecting a ball with the same number of *painted* sticks as the number of *lone pairs* on that atom in your Lewis structure. If you do this correctly, you will find that the number of holes matches the number of bonds that atom has in your Lewis structure. The one exception is the yellow ball, which is reserved exclusively for H atoms because they normally form no more than one bond and are satisfied with a single pair of electrons. The conventional color scheme is yellow or white = H, black = C, red = O, blue = N, green = Cl, orange = Br, and violet = I, but any color ball can be used to represent any atom, as long as you are consistent within a given structure and you choose a ball with the appropriate number of painted sticks.

A single bond is represented by an *unpainted* stick joining two balls. A double bond is represented by two springs joining the two balls, and a triple bond, by three springs. As examples, let us see how one would build models corresponding to the following structures:

$$\mathrm{H{-}\ddot{\underset{\cdot\cdot}{O}}{-}H} \qquad \left[\mathrm{:\ddot{\underset{\cdot\cdot}{O}}{-}H}\right]^{-} \qquad \mathrm{:N{\equiv}N:}$$

The oxygen atom in a water molecule has 2 lone pairs; therefore a red ball (which has 2 red sticks glued into it) should be used. Use an unpainted stick to join each hydrogen atom (a yellow ball) to the oxygen. The oxygen atom in the hydroxide ion, $OH^-$, however, would require a ball with *3* painted sticks glued into it to represent the three lone pairs. It does not matter which color ball you use, as long as it has 3 painted sticks glued to it. However, once you have picked a color, to avoid confusion, you should use the same color if that element occurs again in the same structure.

For $N_2$, which has one lone pair on each N (see the Lewis structure shown above), use 2 blue balls (each with one blue stick glued into it) and three springs to connect them.

The unpainted sticks come in two lengths. Traditionally, the shorter ones are used for bonds to H atoms, but if there is a shortage of sticks, any length will do as long as you are consistent.

# Appendix 11

# Density of Water at Different Temperatures

DENSITY OF WATER AT DIFFERENT TEMPERATURES (g/mL)

| °C | 0.0 | 0.1 | 0.2 | 0.3 | 0.4 | 0.5 | 0.6 | 0.7 | 0.8 | 0.9 |
|---|---|---|---|---|---|---|---|---|---|---|
| 15 | 0.99910 | 0.99909 | 0.99907 | 0.99906 | 0.99904 | 0.99902 | 0.99901 | 0.99899 | 0.99898 | 0.99896 |
| 16 | 0.99894 | 0.99893 | 0.99891 | 0.99890 | 0.99888 | 0.99886 | 0.99885 | 0.99882 | 0.99881 | 0.99879 |
| 17 | 0.99878 | 0.99876 | 0.99874 | 0.99872 | 0.99871 | 0.99869 | 0.99867 | 0.99865 | 0.99863 | 0.99862 |
| 18 | 0.99860 | 0.99858 | 0.99856 | 0.99854 | 0.99852 | 0.99850 | 0.99848 | 0.99847 | 0.99845 | 0.99843 |
| 19 | 0.99841 | 0.99839 | 0.99837 | 0.99835 | 0.99833 | 0.99831 | 0.99829 | 0.99827 | 0.99825 | 0.99823 |
| 20 | 0.99821 | 0.99819 | 0.99816 | 0.99814 | 0.99812 | 0.99810 | 0.99808 | 0.99806 | 0.99804 | 0.99802 |
| 21 | 0.99799 | 0.99797 | 0.99795 | 0.99793 | 0.99791 | 0.99789 | 0.99786 | 0.99784 | 0.99782 | 0.99780 |
| 22 | 0.99777 | 0.99775 | 0.99773 | 0.99770 | 0.99768 | 0.99766 | 0.99764 | 0.99761 | 0.99759 | 0.99756 |
| 23 | 0.99754 | 0.99752 | 0.99749 | 0.99747 | 0.99745 | 0.99742 | 0.99740 | 0.99737 | 0.99735 | 0.99732 |
| 24 | 0.99730 | 0.99727 | 0.99725 | 0.99722 | 0.99720 | 0.99717 | 0.99715 | 0.99712 | 0.99710 | 0.99707 |
| 25 | 0.99705 | 0.99702 | 0.99700 | 0.99697 | 0.99694 | 0.99692 | 0.99689 | 0.99687 | 0.99684 | 0.99681 |
| 26 | 0.99679 | 0.99676 | 0.99673 | 0.99671 | 0.99668 | 0.99665 | 0.99663 | 0.99660 | 0.99657 | 0.99654 |
| 27 | 0.99652 | 0.99649 | 0.99646 | 0.99643 | 0.99641 | 0.99638 | 0.99635 | 0.99632 | 0.99629 | 0.99627 |
| 28 | 0.99624 | 0.99621 | 0.99618 | 0.99615 | 0.99612 | 0.99609 | 0.99607 | 0.99604 | 0.99601 | 0.99598 |
| 29 | 0.99595 | 0.99592 | 0.99589 | 0.99586 | 0.99583 | 0.99580 | 0.99577 | 0.99574 | 0.99571 | 0.99568 |
| 30 | 0.99565 | 0.99562 | 0.99559 | 0.99556 | 0.99553 | 0.99550 | 0.99547 | 0.99544 | 0.99541 | 0.99538 |

To find the density of water at any temperature between 15.0°C and 30.9°C, read the entry where the row containing the whole degree reading intersects the column containing the tenths reading.

For example, at 22.4°C, the density of water is 0.99768 g/mL.

# Appendix 12

# Using a Spectrophotometer

## A. Interaction of Light with Matter

When a molecule absorbs different wavelengths of light, the molecule can respond by changing in some way. Light in the visible or ultraviolet region causes the electrons in the molecule to move into higher energy levels. When light energy is absorbed by a sample of matter, light of different wavelengths is absorbed to differing extents. The fraction of the light absorbed by a sample depends on the sample itself, the wavelength, the concentration of the absorbing species, and the length of the light path through the sample. Suppose you have a sample dissolved in a solvent at a concentration c in a *cuvet*, a sample holder whose inside dimension is the *path length* symbolized by *b*[1]. If light of initial intensity $I_0$, shines onto this sample and some of the light is absorbed by the sample, then the light intensity will be some smaller amount symbolized by *I* when it exits the sample.

For most compounds at most wavelengths, the following relationship holds:

$$\log\left(\frac{I_0}{I}\right) = A = abc \qquad (1)$$

where $A$ = *Absorbance* of the sample under study
$a$ = proportionality constant called the *absorptivity*
$b$ = *path length* of the sample cell[1]
$c$ = *concentration* of sample

A compound that obeys equation (1) is said to be obeying the *Beer-Lambert law* (*Beer's law* for short) at that wavelength. If the compound, wavelength, and path length are all kept constant, then this relation is equivalent to saying that the absorbance and the concentration are directly proportional to one another. As with any two quantities related in this way, a graph of *A* versus *c* will yield a straight line. The slope of the line equals *ab* at the wavelength used. If, as is commonly done, the path length of the sample cell is 1 cm, then the numerical value of the slope is *a*, but its units will still contain the length unit.

---

1. The letter l is also used to symbolize the path length.

The units on $a$ are determined by the units in which $b$ and $c$ are expressed. (The absorbance, being the logarithm of a unitless number, has no units.) For the special case when the concentration is expressed in moles per liter, the absorptivity is generally denoted by ε (Greek epsilon), and is called the *molar absorptivity* of the sample.

To measure $A$, we use an instrument called a *spectrophotometer*. Most spectrophotometers do not measure either $I$ or $I_o$ directly, but they measure and output the ratio $I/I_o$ (referred to as the *transmittance* ($T$) of the sample). The interrelationships between the transmittance and the absorbance of a sample are

$$A = \log\left(\frac{I_o}{I}\right) = \log\left(\frac{1}{T}\right) = -\log(T)$$

Depending on the spectrophotometer used, you may be able to read $T$, $A$, or some combination of these. Since absorbance is directly related to the concentration of the sample, it is generally the most desired of these quantities.

The most important application of Beer's law and spectrophotometry is for determining concentrations of unknown samples, known as a *spectrophotometric determination* of the concentration of the sample. One begins with choosing an appropriate wavelength for the absorbing species by obtaining an *absorption spectrum* (scanning the sample over a range of wavelengths). Generally the wavelength where the absorbance is the maximum ($\lambda_{max}$) is selected as the analytical wavelength. At this wavelength, the absorbance readings of a set of standard solutions of the species at known concentrations are measured, and a calibration curve is plotted for absorbance vs. concentration. Finally, the absorbance of the unknown sample solution is measured (at the same wavelength). The concentration of this unknown sample is then determined from the calibration curve.

## B. Operating a Spectrophotometer

The *spectrophotometer* is one of the more common chemistry laboratory instruments. As with any laboratory equipment, the accuracy and precision of the data collected are affected by whether the operator is using the equipment properly.

1. General Considerations

a. Use and Care of Cuvets

The outside can be dried with a *soft* lint-free tissue or cloth (such as *Kimwipes®* tissues). To avoid scratches, store them in the Styrofoam holders provided. Just prior to insertion into the instrument, the outside should be wiped again to remove fingerprints and other material that might be adhering to the surface. Also check to see that there are no air bubbles in the cuvet. Gently tap the bottom of the cuvets on the bench top to dislodge any large air bubbles from the sides of the cuvet. It is important that soon after use, the cuvets are thoroughly rinsed with deionized water to prevent them from being caked with or stained by the samples.

The cuvets used in CHEM 131L are made of plastic, rectangular in shape, and with a flat bottom. It may be that two of the sides are grooved or frosted, and the other two are clear, or all four sides may be clear. It is important that this type of cuvet is placed into the cell holder in the instrument with one of the clear surfaces facing the incoming beam of light. You may label them with a *small* piece of labeling tape applied to the top of one of the grooved sides. Be sure to remove the labeling tape at the end of the lab period. Cuvets should be filled about ¾ full (to within about ¼" of the top) with the solution being analyzed. *Also, since these are made of plastic, do not rinse them with acetone, which can craze the surface and ruin the cuvet.*

## b. Preparation and Use of a Blank Sample

A *blank* is a sample that contains everything in the unknown sample except the material being analyzed. Often, this is simply water (the solvent), but a blank may contain other substances such as the acid added to adjust the acidity of the sample. If we were interested in the absorbance of some solute dissolved in water, we would not want to include the absorbance of the water itself. By measuring the absorbance of the blank alone, we can subtract this value from the absorbance of the solution to obtain the absorbance of the solute.

In the case where only one wavelength is involved, this can be done simply by inserting the blank into the instrument and setting it to read zero absorbance. All subsequent readings of unknown samples would automatically be corrected for the absorbance of the solvent. This is equivalent to “zeroing” or “taring” a balance to zero.

Appendix 13

# Using Excel®
# for Graphing (Mac)

## A. INTRODUCTION – SOME BASICS ABOUT GRAPHING

A *graph* is essentially a pictorial way of presenting a set of data. This can be done in a variety of ways (such as a circle or "pie," or a bar graph), but the most common type of graph used in science is a *line graph*, where each point in a two-dimensional space is designated by an ordered pair of numbers. By plotting the points of the data set, it is often possible to discover or interpret the relationship between the two quantities being plotted. Conventionally, the axes are drawn perpendicular to one another. The horizontal axis is conventionally called the *abscissa* or *x*-axis, while the vertical axis is conventionally called the *ordinate* or *y*-axis. Conventionally, the *x* coordinate of the point is specified first. The point where the two axes cross (with coordinates (0,0)) is called the *origin* of the graph.

In choosing how to plot the data (i.e., which quantity is plotted on which axis), the convention is to plot the *independent variable* (i.e., the one over which the experimenter has control) on the *x*-axis and the *dependent variable* (the one which is measured in response to changes in the independent variable) on the *y*-axis. This means that changes in the independent variable are thought to be the *cause* of the changes in the dependent variable. Mathematically, we say that the dependent variable *is a function of* the independent variable. If you are asked to "plot A *vs.* B," or "make a graph of A *vs.* B," that means that quantity or variable A should be graphed on the *y*-axis and quantity or variable B on the *x*-axis.

Usually, the points on a graph are connected by some sort of *smooth* curve (which could be a straight line; to a mathematician, *any* line on a graph is called a curve), not a point-to-point, "connect-the-dots" sort of broken line. If this is the case, this indicates the existence of some sort of trend or regular relationship between the variables, as opposed to a group of unrelated random or isolated events. This would allow you to predict the results of an experiment at points between data points (*interpolate*), and to predict the results of an experiment at points beyond the range of the measured data (*extrapolate*).

In this course, *all* graphs submitted as part of a lab assignment must be generated using a computer graphing program. There are many different programs to choose from, but one of the most common is Microsoft Excel®, which is available on the public computers on campus. What is presented here is a very introductory guide to using Excel® (based on Excel® for Mac 2016) to produce graphs. This should give you the basic information and skills necessary to complete assignments in General Chemistry lecture and laboratory. This program is extremely powerful and sophisticated, with capabilities far beyond those required for this course or discussed below. If you haven't used Excel® before, consider this a good first step in learning how to use this program for this purpose.

## B. USING EXCEL® TO PRODUCE A GRAPH

### Entering Data

The most common way to organize data in Excel is by creating a vertical column with a label in the first row.

Example: Solutions of varying concentration are analyzed by an instrument which generates a signal based on the solution concentration.

|   | A | B |
|---|---|---|
| 1 | Concentration (mol/L) | Signal |
| 2 | 1 | 0.009 |
| 3 | 2 | 0.021 |
| 4 | 5 | 0.046 |
| 5 | 10 | 0.095 |
| 6 | 20 | 0.213 |

In this case the independent quantity is Concentration and the dependent variable is Signal (we know the concentration, but we have to measure the signal). This means that Concentration should be on the x-axis and Signal should be on the y-axis. In Excel, the x-axis values need to be to the left of the y-axis values.

## Making a Line Graph ("Scatter Chart")

For starters, Excel refers to graphs as "charts". By graphing this data, we can see the relationship between signal and concentration.

1. Select the data to be graphed by clicking and dragging the cursor over the range of cells containing your data (including labels).
2. Select the **Insert Tab** from the Ribbon.
3. Select the **X Y Scatter button** in the **Charts** group and select the first subtype (Scatter).

| | A | B |
|---|---|---|
| 1 | Concentration (mol/L) | Signal |
| 2 | 1 | 0.009 |
| 3 | 2 | 0.021 |
| 4 | 5 | 0.046 |
| 5 | 10 | 0.095 |
| 6 | 20 | 0.213 |

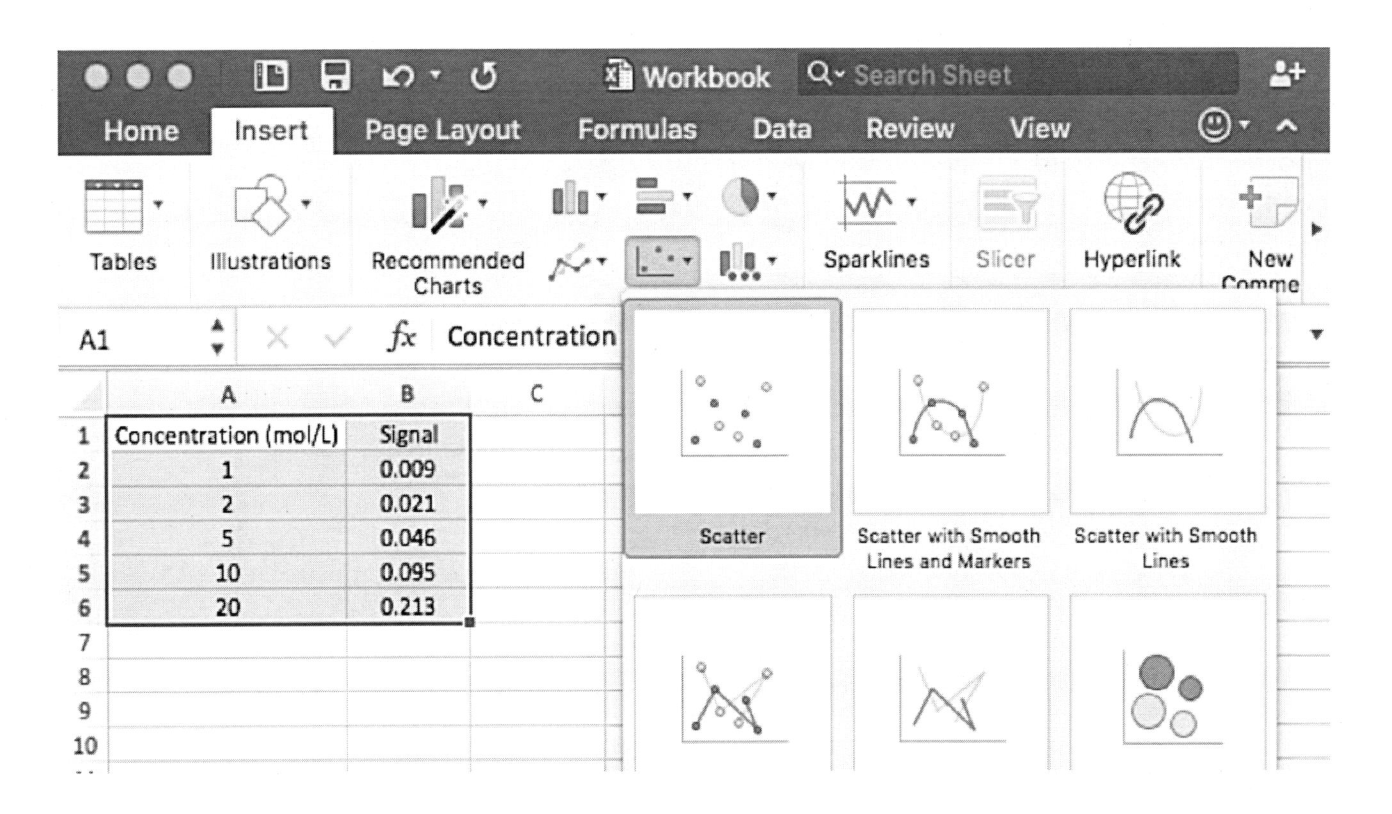

4. If the XY Scatter Chart is covering the data, move it by placing your mouse in a blank area of the chart until your mouse turns into a four headed arrow, hold down the left mouse button and drag the chart to a clear area on the worksheet.

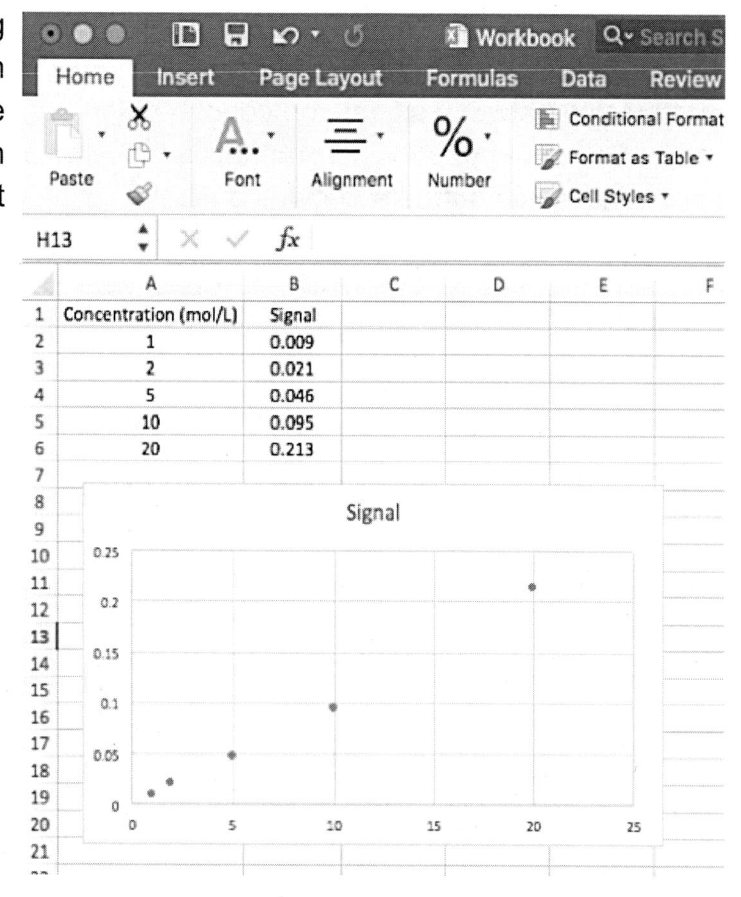


## Changing the Chart Title

1. Select the "**Chart Title**" above the chart itself and type directly over it with your desired title.

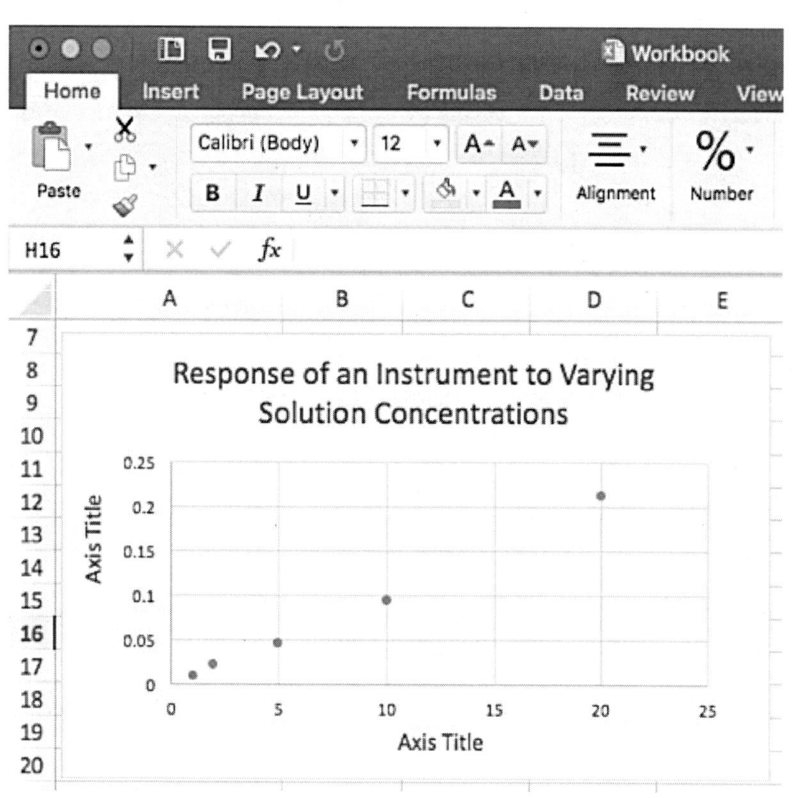

If there is no title:

2. Select **Chart Design** from the Ribbon.
3. Select **Add Chart Element**.
4. Under **Chart Title**, select the **Above Chart** option. The Chart Title text box appears above the chart.
5. Type your chart title and press **ENTER** on your keyboard.

**Note:** Depending on the length and complexity of the title you select, you may have to adjust the size of the overall chart borders, the size of the chart itself, and the relative positions of the chart and the title box. This is done by clicking one edge or corner of the box you wish to size or move and dragging it with the cursor. You can also click the "Format" tab, then use the "Size" group menu for this purpose.

## Adding Axis Titles

Axis labels should include the quantity being measured (data type) and the units (IMPORTANT!).

1. Select **Chart Design** from the Ribbon.
2. Select **Add Chart Element**.
3. Under **Axis Titles**, select **Primary Horizontal** or **Primary Vertical** (you must add each axis title individually).

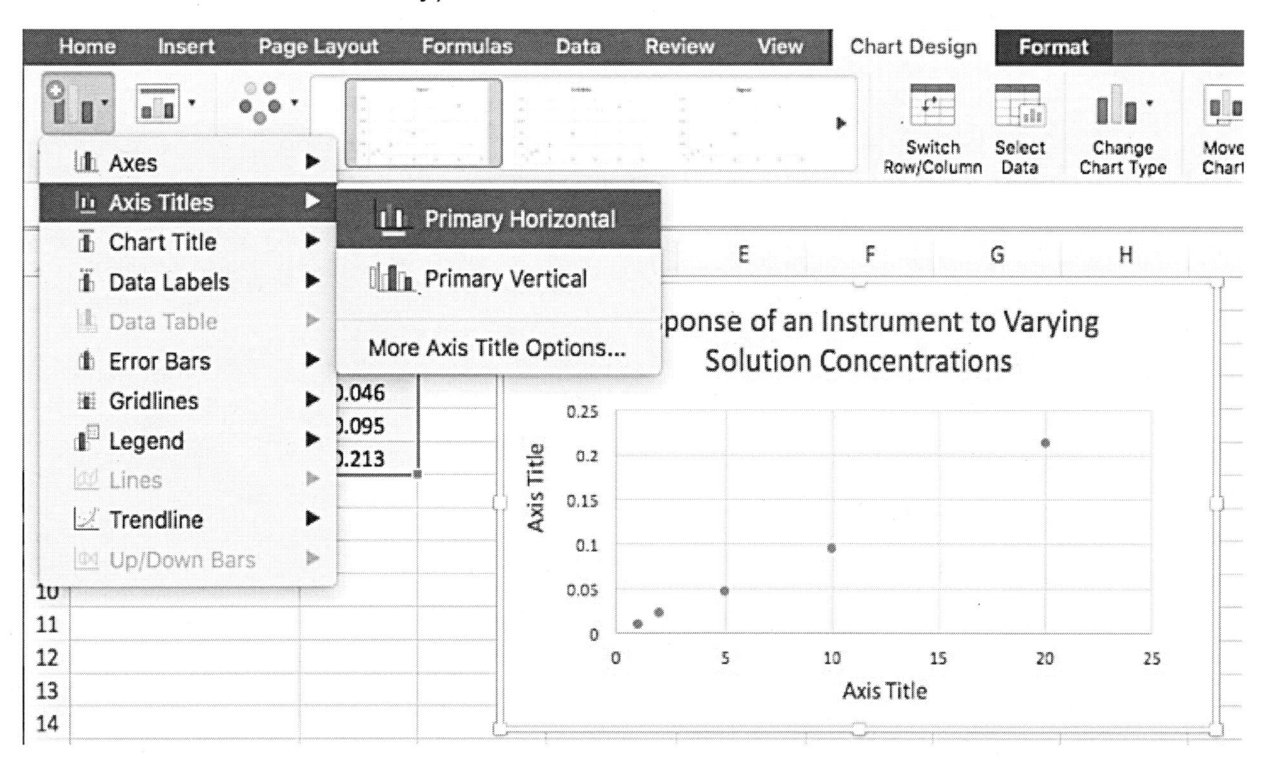


4. Type the vertical axis title you want (Signal (units) was used in this example) and press ENTER.

5. Type the horizontal axis title you want (Concentration (mol/L) was used in this example) and press ENTER.

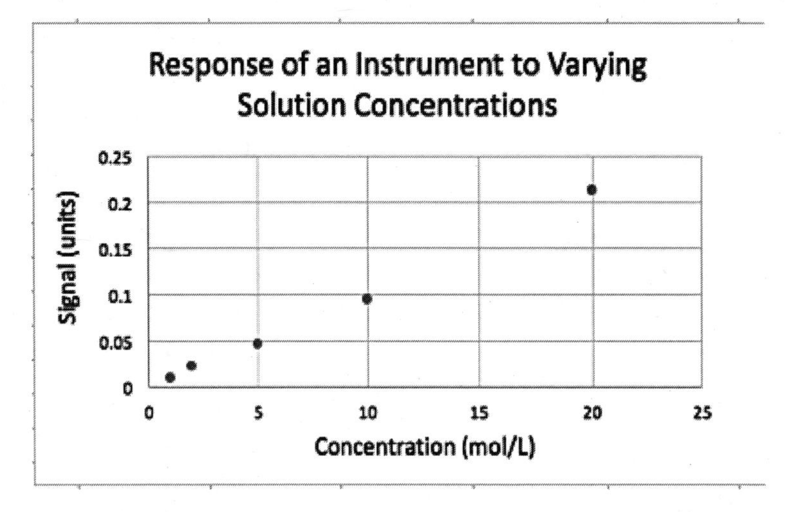


## Best Fit Line, Slope and Intercept

Often, we need to determine the mathematical relationship between the points on a graph. In the graph above, you can see that the points form a straight line. A best fit line can be added to the graph by the following the steps below:

1. Click once on the chart to select it.
2. Select **Chart Design** from the Ribbon.
3. Select **Add Chart Element**.
4. Under **Trendline**, select the **Linear** option. The equation of that line will follow the form: y = mx + b where m is the slope and b is the y-intercept.

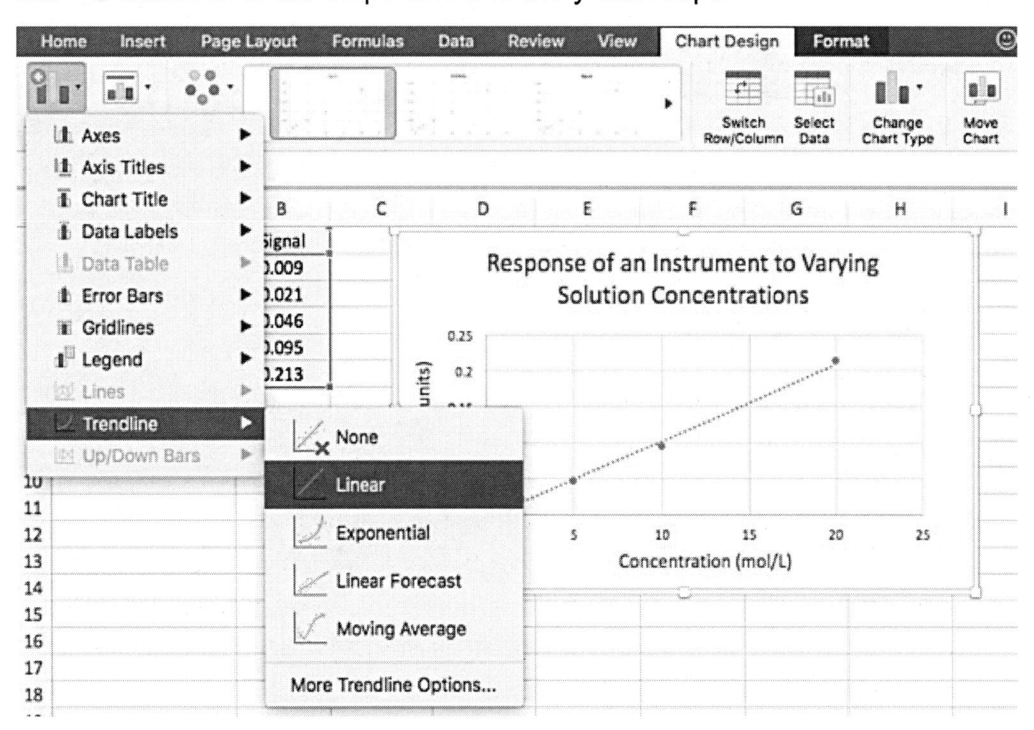

5. Double click on the trendline on the chart.
6. The **Format Trendline** dialog box will appear.
7. At the bottom of this dialog box, check the box for **Display equation on chart**. You may also want to check the box for **Display R-squared value on chart** for some graphs.
8. Click **Close (X) button** in upper right hand corner of the Format Trendline box to close.
9. When you return to your graph, it will have the equation of the best fit line, along with the R-squared value. Often the equation is in the middle of the graph and may be covering data points, so you can move it to one side by dragging the equation.

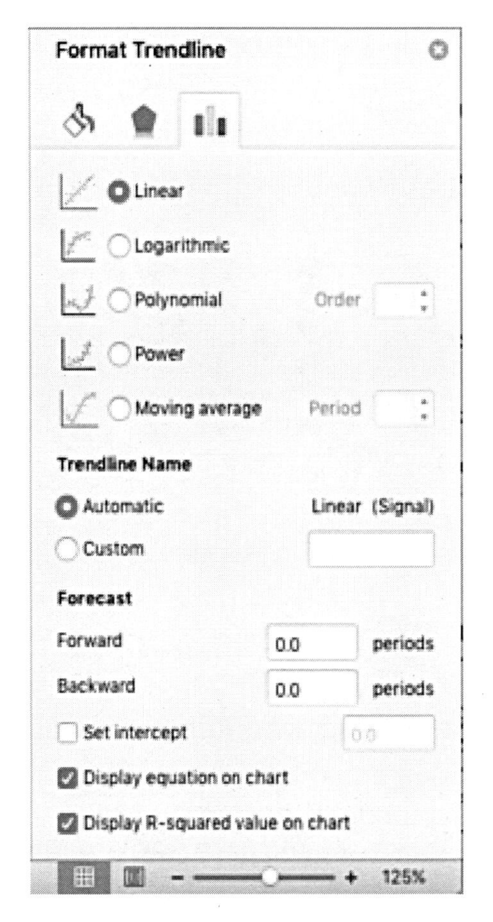


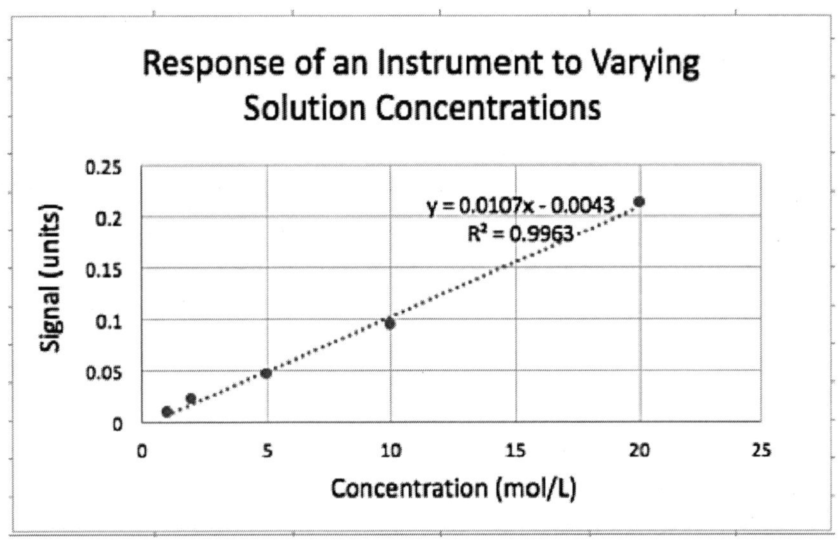


The R-squared statistic is a rough measure of how well the data fit the model of a straight line. The closer this value is to 1.0, the better the fit of the data to the linear model. In this graph, the points are all very close to the best fit line, so the R-squared value is high (above 0.99). An R-squared statistic below 0.9 is typically an indicator of a poor fit, and might prompt you to look for an outlier in the data points.

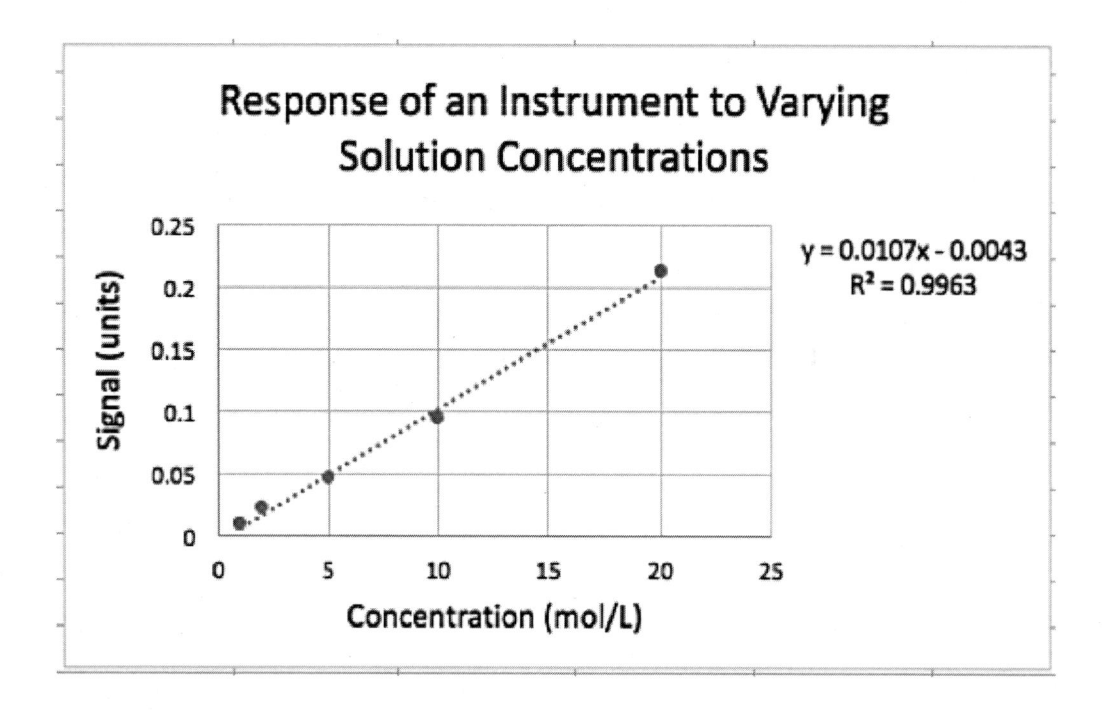


### Fine-Tuning and Printing

When printing, *you must make sure the graph is selected.* When selected, the graph will have a border around it and the 'Chart Tools' become available on the top ribbon. When the graph is not selected these 'Chart Tools' go away.

### Changing numbering on an axis

1. Click once on the chart to select it.
2. Double click on the axis you wish to change.
3. The Format Axis box will appear on the right of the screen. The axis selected (x or y) is indicated by which axis has a box around it on the graph. Click on the other axis to move between them. In this example, the x-axis is selected, and so all of the values in the Format axis box relate to x-axis scale. Often the 'Auto' function does a good job of selecting the **Bounds** (range of the axis) and **Units** (increments where numbers are displayed for axis). You can type over any of the values and hit enter to change the way the data are displayed.

### Gridlines

1. Click once on the chart to select it.
2. Double click on the gridlines on the graph. The Format Major Gridlines box will appear on the right of the screen.
3. Click on the paint can icon. Turn gridlines on and off by selecting either "no line" or "automatic". Toggle between horizontal and vertical gridlines by selecting the respective set of gridlines on the graph.

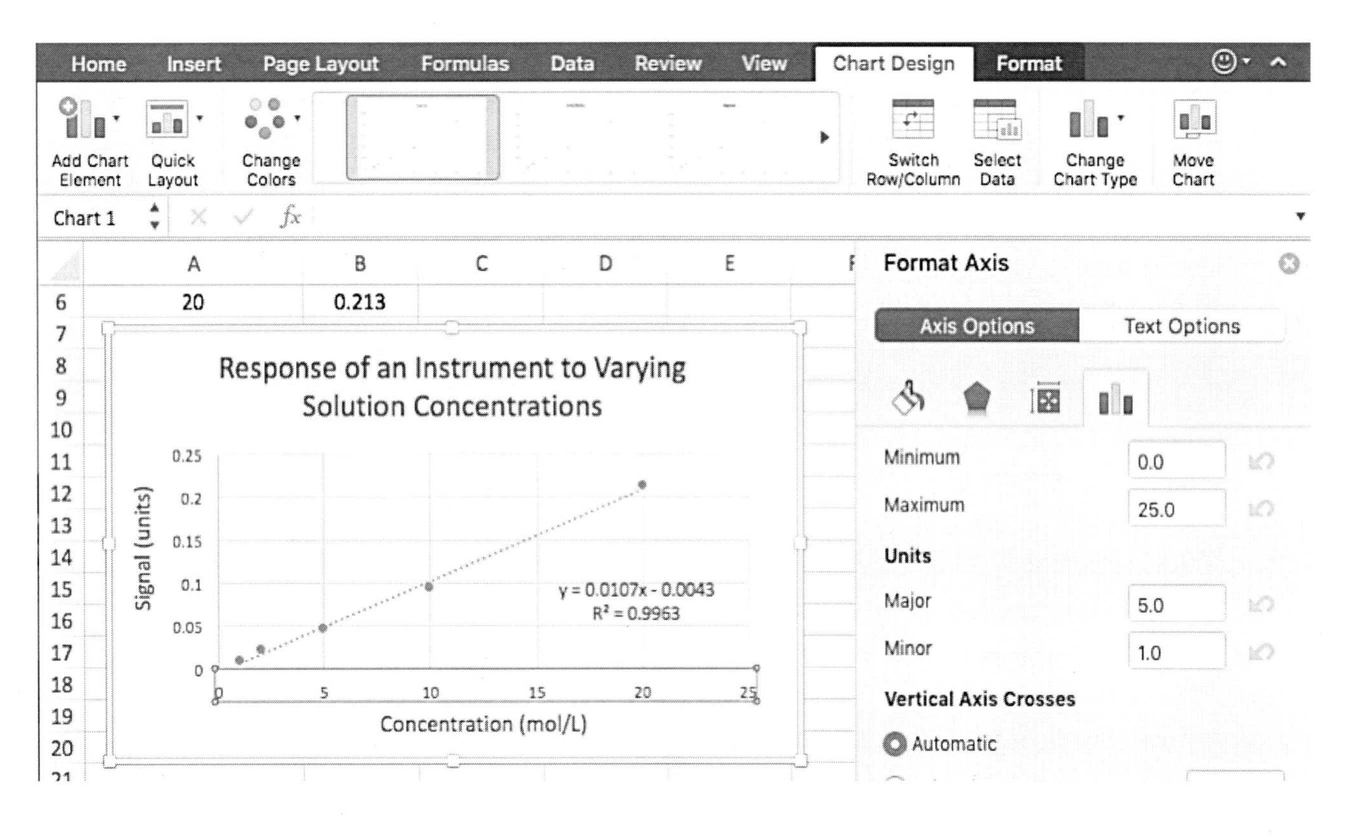
Home
Insert
Page Layout
Formulas
Data
Review
View
Chart Design
Format
Add Chart Element
Quick Layout
Change Colors
Switch Row/Column
Select Data
Change Chart Type
Move Chart
Chart 1
Format Axis
Axis Options
Text Options
Minimum
0.0
Maximum
25.0
Units
Major
5.0
Minor
1.0
Vertical Axis Crosses
Automatic
Response of an Instrument to Varying Solution Concentrations
Signal (units)
Concentration (mol/L)
y = 0.0107x - 0.0043
R² = 0.9963
20
0.213

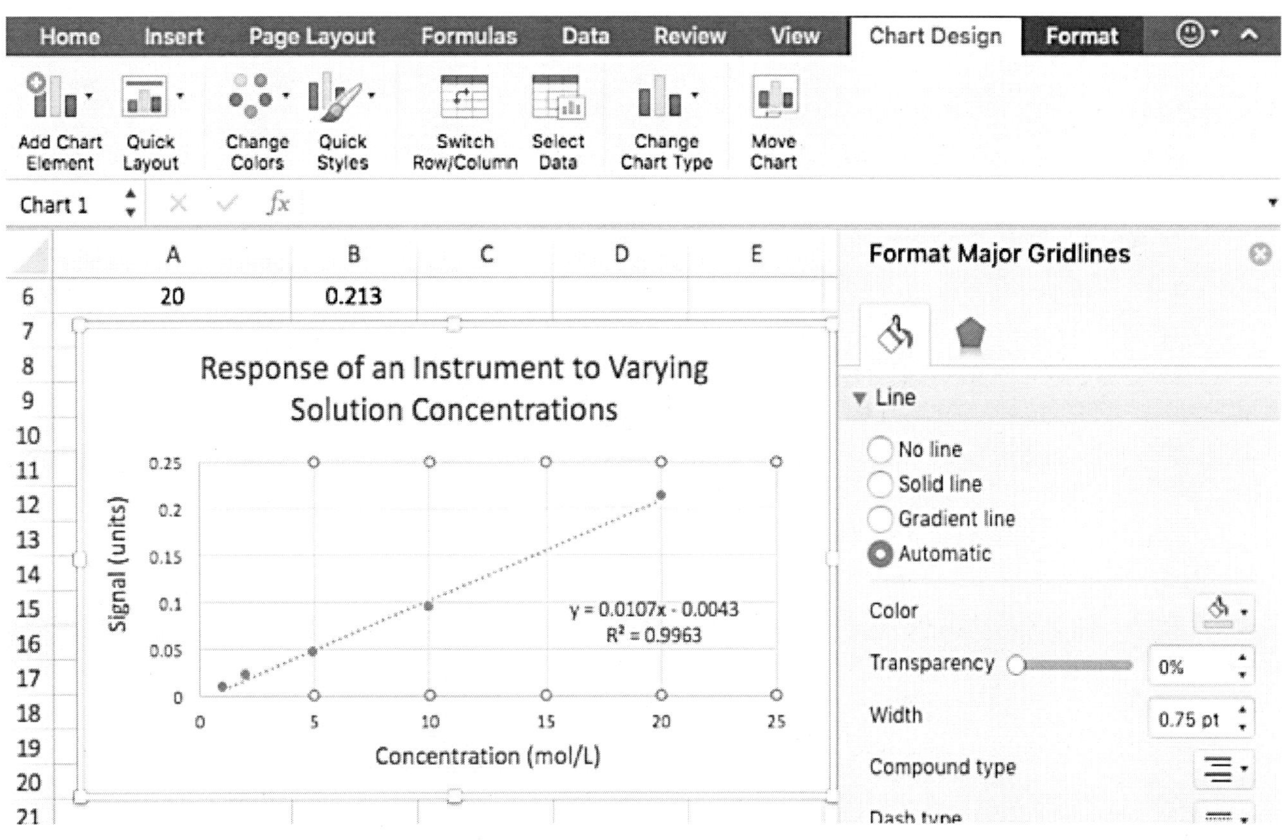
Home
Insert
Page Layout
Formulas
Data
Review
View
Chart Design
Format
Add Chart Element
Quick Layout
Change Colors
Quick Styles
Switch Row/Column
Select Data
Change Chart Type
Move Chart
Chart 1
Format Major Gridlines
Line
No line
Solid line
Gradient line
Automatic
Color
Transparency
0%
Width
0.75 pt
Compound type
Response of an Instrument to Varying Solution Concentrations
Signal (units)
Concentration (mol/L)
y = 0.0107x - 0.0043
R² = 0.9963
20
0.213

### Printing

1. To print the graph alone, on its own page, select it on the worksheet.
2. Click the **File** tab on the Mac's Ribbon (above the open Excel workbook).
3. Select **Print**.
4. The Print dialog box will open. You can switch the orientation between Portrait and Landscape based on what's visually appropriate for your graph.
5. Make your selections and click **Print**.

## C. CHECKLIST FOR A PRESENTABLE GRAPH

The following items are required for a good graph (usually, content may vary).

1. Useful title
   a. More than just "Y vs. X"
   b. Not just "Concentration", but "Concentration of NaCl in..."
2. X and Y axis labels (**with units in parentheses**)
3. Appropriate numerical scale on x and y axes
   c. Conventional increments count by 1, 5, 10, 50 or 100, etc... (or 0.001, 0.005, 0.0025, etc...)
   d. Make sure all data are on the graph
   e. If a point has a y-value of 102 and the y axis only goes to 100, the point will not be visible
   f. Do not include extra decimals in the axis labels
   g. If the range is 0 to 200, the numbers should be in the format 0, 100, 200 and not in the format 100.000, 200.000, etc...
4. When printing, choose landscape or portrait depending on what looks best for the graph